Virtual Coach,
Virtual Mentor

Virtual Coach,
Virtual Mentor

David Clutterbuck
*Sheffield Hallam and
Oxford Brookes Universities*

Zulfi Hussain

≡*IAP*

INFORMATION AGE PUBLISHING, INC.
Charlotte, NC • www.infoagepub.com

Library of Congress Cataloging-in-Publication Data

Virtual coach, virtual mentor / [edited by] David Clutterbuck Zulfi Hussain.
    p. cm.
  Includes bibliographical references.
   ISBN 978-1-60752-308-6 (pbk.) – ISBN 978-1-60752-309-3 (hbk.) – ISBN
978-1-60752-310-9 (e-book)
1. Mentoring. 2. Mentoring in business. 3. Mentoring in education. 4.
Telematics–Social aspects. I. Clutterbuck, David. II. Hussain, Zulfi.
  BF637.M45V57 2009
  658.3'124–dc22

                                                              2009039169

Printed in the United States of America

CONTENTS

SECTION I

WHAT DO WE KNOW ABOUT VIRTUAL COACHING AND MENTORING?

SECTION II

ORGANIZATIONAL CASE STUDIES

SECTION III

INDIVIDUAL CASE STUDIES

FOREWORD

This book has come about from an evolution of our thinking over a number of years. One of us (David) had been deeply engaged in coaching and mentoring since the concept of formal, structured relationships emerged in the early 1980s. Inevitably, these relationships revolved around face to face meetings, sometimes interspersed wijh occasional letters or phone calls. The other (Zulfi) came to coaching and mentoring through his work at the cuttingedge of telecommunications applications. Our experiences converged when British Telecom initiated one of the first cross-border e-mentoring programmes. The success of this programme started a journey of personal experimentation that has convinced us both that virtual coaching and mentoring are very powerful, effective developmental tools—different but equal to face to face relationships. We've also come to recognise that the many different media available to assist the coaching or mentoring dialogue can be even more powerful when used together.

We have designed the book to provide an overview of both current theory and current knowledge of good practice. In addition, case studies—both of programs and individual experiences—give insights into how to get the best out of the virtual learning partnership. We hope you enjoy the insights we provide!

—**David Clutterbuck & Zulfi Hussain**

INTRODUCTION

In choosing topics, around which to construct a book, we always start with a series of four critical questions:

- How well is this subject already covered?
- Is there something new to say about it?
- Can a new book make a difference in how people work, learn or behave?
- Who cares?

In the case of virtual coaching and mentoring (or e-mentoring and e-coaching; or coaching/ mentoring by wire—choose your own preferred nomenclature!) there are hundreds, perhaps thousands of programmes and initiatives across the world. Yet there is comparatively little in the way of comparison of good practice or academic evaluation of what does and doesn't work. We found numerous individual case studies, but a dearth of empirical research and no significant collection of cases to illustrate the diversity of applications. So the answer to our first question was no.

Just because something is not well covered, it doesn't mean there is anything interesting or useful to say about it at book length. We wanted to do more than simply recount our own experience of virtual coaching/mentoring schemes. We were heartened by a number of thoughtful recent academic studies and different ways of looking at the e-learning interaction.

Gathering case examples from different countries and environments convinced us that there was potential for significant cross-learning from the different approaches adopted.

Virtual Coach, Virtual Mentor, pages ix–xi
Copyright © 2010 by Information Age Publishing

And who cared? We decided that we did, because e-mentoring has the potential to reach enrich the lives of far more people, far more rapidly, than any other mechanism we are aware of. For example, one of us has the ambition to boost the economies of developing countries by creating large-scale e-coaching and e-mentoring programmes. An essential part of this vision is that learning will be two-way—the contract in each relationship will include developing an understanding and appreciation of each other's culture.

HOW THE BOOK WORKS

The structure of the book is straightforward. We start with three initial chapters, which set the scene in a readily accessible way. Section I provides a summary from around the world of what is known on this topic from academic and practitioner literature. Chapter 1 provides an overview of the topic. It provides background and takes an initial look at the pluses and minuses of various media in supporting the learning conversation and relationship. Chapter 2 looks at some of the practicalities of establishing and sustaining an effective virtual coaching or mentoring programme, and of making the learning relationship work. Chapter 3 sketches out the main technologies involved and how we can expect them to develop. We include telephone coaching and mentoring on the grounds that just over 100 years still makes it a newcomer in the arsenal of communication methods, and because it forms one of the big three coaching and mentoring media (the other two being face-to-face conversations and email). First we describe the various media currently available for virtual coaching and mentoring, looking at the practicalities of using them from both a programme and participant perspective. Then we review various ways of integrating existing media and point towards emerging media, which may influence mentoring and coaching in the next decade.

Next we take a more academic view with five chapters addressing what has been learned from the relatively limited amount of research into virtual coaching and mentoring. Paul Stokes reviews practice in the UK and Europe and in particular, lessons from two e-mentoring programmes, one for small business owners, one in the context of a university. Kim Rickard provides an antipodean view of e-mentoring for small business. Randy Emelo and Tom McGee from the United States examine e-mentoring from the perspective of a structured delivery system, providing practical examples of how the technology works. Lisa Boyce and Gina Hernez-Broome return the focus to coaching rather than mentoring, in the context of leadership development.

In Sections II and III, we present a series of case studies from a variety of applications and locations around the world. The first group of cases relates to programmes of e-coaching or e-mentoring. We begin with a detailed case study from Finland, based upon a community of interest within academia.

This is followed by the story of the Brightside Trust, a UK-based organisation that creates and manages community-based e-mentoring programmes. Several of these programmes are described in detail. Link to Life is a pilot programme to offer online coaching support to trainee teachers. AskMax, by contrast, is an e-mentoring system designed to support busy executives and managers within large organisations. Finally, Janet Wright and Jean Simpson offer insights into e-mentoring in a science faculty.

The second group (Section III) describes individual e-relationships from several countries and the lessons learned from them. We also return to the topic of supervision with Julie Hay's case study of a telephone-based supervisory relationship.

We feel that *Virtual Coach, Virtual Mentor* provides a wide variety of perspectives on a rapidly growing phenomenon. We hope and intend that it should make a timely and significant contribution to good practice and to encouraging more practitioners, their clients and more organisations to experiment with using electronic media to enrich coaching and mentoring.

Our view of e-coaching and e-mentoring is firmly one that these new media are less a replacement for traditional face-to-face than an enhancement of learning alliances in general. We see no evidence of fewer face-to-face coaching or mentoring relationships—on the contrary, they continue to become more popular and widespread. Rather, we see that virtual coaching and mentoring both enrich predominantly face-to-face relationships, by connecting partners at times between formal meetings, and open up coaching and mentoring to new audiences and new applications.

Nations, communities and organisations in the 21st century are increasingly dependent on the quality of dialogue that takes place between individuals and groups. Lack of dialogue leads to inadequate sense of purpose, conflict and antagonism, reinforcement of dysfunctional fixed views and a tendency for change to occur drastically and disruptively, rather than as a gradual assimilation of new knowledge, perspectives and capabilities. Whatever the context, coaching and mentoring are the two most powerful methods we have to help people learn and become comfortable with learning dialogue. Learning how to coach and mentoring using the new media, or a mixture of new and old media, not only brings all the traditional benefits of the learning alliance, but has the potential to contribute significantly to the creation and sustenance of flexible, compassionate, healthy corporate and societal environments.

—**David Clutterbuck & Zulfi Hussain**

SECTION I

WHAT DO WE KNOW ABOUT VIRTUAL COACHING
AND MENTORING?

CHAPTER 1

WELCOME TO THE WORLD OF VIRTUAL COACHING AND MENTORING

David Clutterbuck

WHAT DO WE MEAN BY VIRTUAL COACHING AND MENTORING?

We have found a number of definitions, each with its own emphasis, dependent on the type of program. We have also found some more generic definitions, such as this one from the U.S. National Mentoring Working Group, set up by the United Way of America:
    E-mentoring:

- Is a caring, structured relationship
- Focuses on the needs of the mentored participants
- Adds value to the lives of those involved
- Uses technology to connect people across time and/or distance

A more ponderous definition comes from Single and Muller (2001, p. 108): E-mentoring is "a relationship that is established between a more senior individual (mentor) and a lesser skilled or experienced individual

Virtual Coach, Virtual Mentor, pages 3–29
Copyright © 2010 by Information Age Publishing

(protégé),[1] primarily using electronic communications, that is intended to develop and grow the skills, knowledge, confidence, and cultural understanding of the protégé to help him or her succeed, whilst also assisting in the development of the mentor." Other authors echo similar sentiments (e.g., Perren, 2002), but as Stokes et al. (2003) point out, all this amounts to is traditional mentoring with a bit of "e" thrown in.

Similar problems arise with defining e-coaching. Using Google, we searched a range of websites offering e-coaching services. Typical was the statement: "E-coaching is a way of providing executive coaching services with remote communication." Another site saw e-coaching as online therapy. The general theme was "doing what I've always done, but online."

Bierema and Merriam (2002) recognize this limitation in their definition of e-mentoring: "A computer-mediated, mutually beneficial relationship between a mentor and a protégé, which provides learning, advising, encouraging, promoting and modelling, that is often boundaryless, egalitarian and qualitatively different than traditional face-to-face mentoring" (p. 212).

This still seems something of a cop-out, however, in that it merely asserts a qualitative difference without explaining what it is. All the behaviors listed are common to functions of face-to-face sponsorship mentoring.

Ensher and Murphy (2007) define e-mentoring slightly more widely (from a technological perspective) as "a mutually beneficial relationship between a mentor and a protégé, which provides new learning as well as career support, primarily through email and other electronic means (e.g., instant messaging, chat rooms, social networking spaces, etc)" (p. 300).

The simplest definition we have been able to devise, which covers both e-coaching and e-mentoring is: *A developmental partnership, in which all or most of the learning dialogue takes place using email, either as the sole medium or supplemented by other media.*

These generic definitions to some extent obscure the significant variety in the nature, scope and process of the developmental partnership. The relationship may be:

- For a specific or broad purpose (e.g., learning a particular skill or developing a clearer sense of career direction)
- Synchronous or asynchronous (i.e., in real time or with time lapses between responses)
- One-to-one (the most common) or one-to-group
- Formal or informal (i.e., as part of a structured programme or simply between people, who come together naturally)
- Relatively directive or relatively non-directive (different models of coaching and mentoring predispose the relationship to differences in style and interpersonal dynamics)

This short list already identifies one significant difference between coaching/mentoring and their e-equivalents. Face-to-face and telephone coaching are by their very nature synchronous; with virtual developmental relationships, you have a choice. Other critical differences, which we will examine below in the context of pluses and minuses of different approaches, include:

- Equalization of power. Some models of mentoring and coaching have a dynamic, in which the more senior person's authority and seniority may affect the conduct and openness of the conversation significantly. Email conversations tend to be much more power-neutral, not least because they strip out visual clues and symbols (Hamilton & Scandura, 2003).
- The quality and comprehensiveness of the record of conversation. Taking notes in face-to-face dialogue is highly disruptive of attentiveness.
- What Stokes refers to in Chapter 4 as the "leanness" of text-based communication pushes participants to concentrate on content of the conversation, rather than on style
- In face-to-face and telephone dialogue, reflective thinking time has to be created; while in virtual learning alliances, it is automatically present, even during synchronous communication.

Ensher and Murphy (2007) qualify their definition with two important differentiators. Firstly, "these relationships typically cut across internal and external organisational boundaries as well as geographic and time zones and have the capacity to be more egalitarian than face-to-face" (p. 300). In practice, many in-company e-mentoring and e-coaching programs retain the organzational boundaries (commercial confidentiality being one of the influencing factors); moreover, many face-to-face programs are designed to bridge the boundary between an organization and its external communities. (For example, the UK's learning mentor program uses company volunteers to work with schoolchildren.)

However, the capacity for e-mentoring to expand beyond normal boundaries is much greater. Mentoring across geographical and time zones face-to-face is rare because of the practical difficulties. Email does, by and large, make it easier to strip out the power symbols from communications. However, they can still be maintained in subtle ways. For example, one senior manager in a European company restricts email replies to one or two lines at most, but expects detail from people contacting him. The implication, of course, is that "I am too important to spend anything but the briefest time replying to you."

Their second differentiator is that e-mentoring "has particular utility for various other forms of mentoring, such as peer mentoring, group mentoring, and reverse mentoring." While the authors have an instinctive empathy with this statement, especially with regard to group mentoring (where the technology supports more frequent and more time-convenient interaction), the evidence does not appear to be strong in the case of reverse mentoring. Indeed, the following comment[2] from a manager in a reverse mentoring program suggests a much more complex picture: "We'd talked about stereotypes and the impact of loose language several times by email. I'd got the message intellectually, but it was only when I saw the physical changes in him [as he recounted a recent incident], that it struck home. When I re-read the emails, the anger and resentment were there—I just hadn't given them enough weight. But I hadn't sensed the self-shame and his sadness that the other person [in the incident] hadn't *wanted* to understand the impact they were having on him. That was the point I realized that if I wanted to make changes happen in the organization, I needed to make changes in myself."

Nonetheless, there is enough sense of difference here to make a good case that coaching and e-coaching, mentoring and e-mentoring are substantially different kinds of developmental relationship. Equally, there are very significant differences between the typical conversation that takes place through email and the creative dialogue of e-coaching and e-mentoring. Rather than the technology being simply an add-on to an established process, technology and coaching/mentoring have synthesised into something new and exciting.

Before we examine in more detail what makes e-coaching and e-mentoring (sometimes also called tele-mentoring/tele-coaching and online mentoring/coaching) so different from their traditional counterparts, it may be helpful to distinguish between coaching and mentoring themselves.

E-Coaching versus E-Mentoring: Is There a Difference?

We've found examples of people using these terms interchangeably and in some cases with precisely the opposite meanings. Conclusion: there's a lot of confusion out there. It is possible, however to pull together some distinctions between coaching and mentoring in general, which also apply to their virtual versions. The broad consensus—supported by the history of how models and concepts of coaching and mentoring have evolved in the past three decades—suggests that coaching always has a strong element of performance management. There are specific goals (though recent research—Megginson, 2006—indicates that concentrating on goals too much is detrimental) and measures of whether those goals have been achieved. Mentoring, by contrast, is usually much more holistic in its purpose. As a result, mentoring relationships tend to last longer than coaching relationships.

Further differentiating factors include:

- Coaching can and often does take place within the line (from manager to direct reports, although there are significant conflicts in this arrangement, not least that the line manager may be part of the performance problem! (Ferrar, 2006). Mentoring is usually regarded as a relationship that has to occur outside the line, in order to offer different perspectives. The validity of many largely U.S. studies of mentoring has been undermined by a failure to distinguish between line manager and off-line helping relationships.
- Mentoring includes some functions not normally seen within coaching—for example, helping the individual enlarge their networks, by making introductions. Early U.S. models of mentoring placed significant emphasis on sponsorship behaviours by the mentor—a concept that did not work in a European context and gave rise to the European model, often described as developmental mentoring.

There are some spurious distinctions between coaching and mentoring, often touted in articles, where someone is trying to carve out or seize some kind of moral or intellectual high ground for their particular style of coaching or mentoring practice. Most common is that mentoring is directive and coaching is non-directive, or vice versa. In reality, there are directive and non-directive styles and models of both. Hence the definition used in one of our case studies (Chapter 13) is in our view far too limiting on the nature of the relationship and its potential for mutual learning: *"E-mentoring is a relationship that is established between a more senior individual (mentor) and a lesser skilled or experienced individual (mentee), primarily using electronic communications, and is intended to develop and grow the skills, knowledge, confidence, and cultural understanding of the mentee to help him or her succeed."*

The unifying factors behind coaching and mentoring include:

- They are both forms of learning conversations, which achieve most when they tend towards dialogue.
- They both require reflection: before, during and after the learning conversations. They take people on a journey through personal reflective space (Clutterbuck & Megginson, 1999).
- They are both, in their non-directive forms, a powerful opportunity for mutual learning and, indeed, the mutuality of learning is an indicator of relationship efficacy.
- They both tend to take place within the context of an organization or profession—a working environment, social network, or community of learning. Interestingly, one of us has recently carried out an extensive longitudinal study of mentoring pairs and found that the

organization context does not significantly influence the conduct of the relationship. The learning partnership becomes a cocoon within the organization and succeeds or fails primarily upon the quality of what happens within (Clutterbuck, 2007).

Benefits of Virtual Coaching and Mentoring

Perhaps not surprisingly, most of the research into benefits of coaching and mentoring relate to face to face relationships. Some U.S. studies conflate face-to-face and telephone coaching, so it's difficult to draw any real conclusions about the latter. Indeed, reliable data on the benefits of e-coaching and e-mentoring is very scarce, whether it relates to outcomes for participants or the sponsoring organisation. However, research by Triple Creek (Emelo, 2008; see Chapter 6), an online mentoring provider, found that:

- 85% of respondents reported a positive connection between mentoring and factors that impact retention.
- 81% reported a positive connection between mentoring and factors that impact productivity.
- 77% reported positive satisfaction ratings with their relationships.
- For participants who invested at least one hour per month on mentoring, the above results were even more positive.

Applications of E-Mentoring/E-Coaching

Virtual mentoring and coaching programs can be seen in most of the areas where face-to-face relationships occur as well as in some additional areas. There isn't a broadly accepted classification of mentoring or coaching applications, but one that works reasonably well is community, educational, business/employment and professional. There are some overlaps between these categories, but for the most part, they form separate worlds. These worlds do not necessarily communicate very well with each other. They tend to be represented and championed by different bodies, to use technology differently, and to engage in almost no cross-category research.

Community Oriented
Community mentoring and coaching address a wide range of social needs, such as:

- Integrating immigrants and asylum-seekers into their host societies
- Supporting teenage mothers

- Supporting people who suffer chronic disease, or who require long-term care, or who are on the autistic spectrum
- Helping ex-offenders find employment and "go straight"
- Helping prisoners address issues such as dyslexia
- Preventing young people from disadvantaged inner city backgrounds from going off the rails

Some community applications of virtual mentoring/coaching pose special problems of access or protection, which may limit take-up and would not normally be an issue in face-to-face applications. For example, prisons are generally wary of giving inmates confidential access to computers, for fear of abuse, so more expensive dedicated software resources may be needed. Programs aimed at vulnerable people (those under the age of consent, or with a learning disability, for example) tend for the most part to require robust systems of monitoring conversations (raising issues of privacy that would be seen as intrusive in other contexts). Some, such as Horsesmouth, maintain the anonymity of the participants as an additional safety measure. Horsesmouth is a good example of e-mentoring as a social network, tapping into the altruistic instinct of people to help each other. This instinct works even when the two parties know little or nothing about each other.

Education Oriented

It is not surprising that education was one of the first areas to embrace structured coaching and mentoring. Equally, given widespread experience of distance-learning and web-based learning management systems in education, it has been a logical step to electronic media for coaching and mentoring.

Some of the applications we have found, which may be either face-to-face or virtual, include:

- Helping black students remain at university
- Helping new university faculty acquire tenure
- Supporting schoolchildren, who have literacy and numeracy problems
- Teacher education
- Cross-sector mentoring between head teachers and business people

The same issues arise in education with regard to dealing with vulnerable and under-age persons as in community applications.

Business and Employment Oriented

Structured mentoring and coaching have blossomed in the world of work. Examples include:

- Supporting diversity and equal opportunities objectives

- Supporting people through major transitions (e.g., joining an organization, being promoted to a higher level of management, becoming a director, or retiring)
- Reinforcing major culture change
- Improving staff retention
- Improving the survival rate of small businesses and business start-ups

Small business owners have been targeted in e-mentoring programs in numerous countries, including the U.S., the UK, Eire, and Australia. One of the main reasons small businesses fail is that the owner puts all of his or her effort into growing the business, leaving insufficient time and energy for growing himself. The e-mentor provides a balance, provoking them to think about wider issues, but also being available to assist with guidance when crises occur.

Charities and non-governmental agencies have staff scattered around the developing world. Bringing expatriates back to their home country for training or taking training to them is simply not affordable. It's also very lonely and isolated being in a small office in a remote location. E-coaching and e-mentoring fulfill two key roles in these circumstances—continuous education and social networking—both of which are vital to the effective functioning of field forces.

Multinational companies are increasingly finding that expatriates in senior roles benefit from having a local coach or mentor, who helps them adjust to the culture in which they are working.

The first programs to address maternity mentoring or coaching go back to the early 1990s, but many modern versions now incorporate a large element of email conversation, or are email based. The relationship is typically established six months or so before the intended return to work, with the mentor being another mother who has already made the same transition back into work. Email fits the process well, because the mother does not have to make special arrangements to leave the baby while she comes into the workplace, or disrupt the workplace by bringing the baby in for meetings. Messages from the mentor tend to be colored with what is happening in the workplace at the moment, helping the returning mother feel part of it although she is not there. This kind of e-mentoring requires structure, particularly in the final three months, to ensure that the mother is aware of all the issues she will face and deals with them ahead of the return, as much as possible. Recommended practice is to open out aspects of the email conversations to other key work colleagues in the final month, with the mentor ensuring that they are engaged in the return process. Many returning mothers report that they feel much more welcome and that colleagues are more prepared for them than would normally be the case.

Profession Oriented

Informal mentoring has traditionally been a characteristic of professions, so it is not surprising that professions have taken to virtual mentoring and virtual coaching. The case study of Empathy-Edge is a good example of a general professional program, but others are much more specific to professions. For example, new lawyers benefit from building networks of coaching and mentoring relationships with more experienced colleagues internationally. Clerics, too, find value in e-mentoring. ASK is a U.S. program based on the premise that Christian pastors "as humans . . . have needs to be met [and] are unable to share with anyone, given the nature of their responsibilities and the need for confidentiality." E-mentors help them by giving guidance and occasionally therapeutic help and "enhancing their shepherding skills through long-distance interactive media."

Another U.S. program, MentorNet, focuses on linking women university students at both undergraduate and graduate levesl to mentors in large technology employer organizations.

Benefits and Disadvantages of Virtual Coaching/Mentoring

As we have already discussed, what virtual coaching or mentoring means in a particular circumstance can differ considerably. So in making comparisons with face-to-face mentoring and, say, telephone-based mentoring, it's important to take into account the intent and nature of the relationship.

One of the editors (Clutterbuck) came to e-mentoring and e-coaching initially with a great deal of skepticism about its efficacy. How could you possibly equal the immediacy and energy of a face-to-face conversation with an online conversation? However, interviews with participants in virtual coaching and mentoring programs, along with personal experience of virtual developmental partnerships, produced a gradual but fundamental reassessment. Our view now is that both face-to-face and virtual approaches have great strengths and some weaknesses and that judgments about efficacy need to be rooted in the context of the individual relationship, rather than in general comparisons of one process versus another.

The idea that face-to-face interactions are ideal and that other media are less than ideal stems, according to Harrington (1999), from a preference for what we are used to. In her review of the literature, she concludes that there is

> [There is] no obvious reason why text-based computer mediated communication should necessarily be a poorer medium than face-to-face communication for conveying social information. It may hinder expressively rich communica-

tion, but need not entirely prohibit it. With sufficient time, effort and attention to the task, it is perfectly possible to pack text full with social meaning. Indeed, 'the world of literature makes clear that face-to-face interaction has no inherent advantage over text in this respect.' (Reid, Mallinek, Stott, & Evans, 1996, p. 1018, quoted in Harrington, 1999)

Media richness is said to be influenced by a number of factors. In the context of email communication, the most significant are:

- Capability for immediate feedback
- The number of channels (e.g., tone, facial expression, gestures)
- How personalized the interaction feels
- The extent of language variety

While email communication could be predicted to provide less of all of these characteristics, there is evidence that, in fact, people find ways to compensate for them. Senior managers use email far more frequently than the media richness theories would predict (Markus, 1994), for example. It seems that the more comfortable people become with computer communication, the richer their communication becomes. (One wonders whether, when our prehistoric ancestors first added vocal signals to physical gestures, wise heads shook and thought "this speech idea will never be as effective as the real thing!")

The elements of media richness have a darker side that helps this process. Immediate feedback is not always a positive thing. Face-to-face communication can be relatively shallow and unconsidered, because it creates a *need* to respond quickly. Long pauses for reflection are seen as rude or socially inept, at least in most Western societies. While multiple channels are valuable in checking whether assumptions based on one are born out by others (for example, dissonance between the message and facial expression or posture), the sheer volume of information can be distracting. Indeed, the brain functions by filtering out most of this information from our conscious awareness. Personalization brings with it the trappings of social status and organizational baggage. Given that one of the key functions of an effective mentor is to provide an independent, unbiased perspective, an element of impersonalization in email mentoring is potentially an advantage. (Although a different kind of personalization can be built up as email correspondents build rapport and confidence in each other; it can be argued that the dialogue itself gradually assumes a personality.) Language variety in face-to-face speech includes a great deal of redundancy—for example, hesitations, repetitions and half-finished thoughts—and this can also be distracting, making it more difficult to achieve real depth of dialogue.

A critical factor in how effectively a coaching/mentoring pair use any medium is how comfortable they feel with it. Comfort in this case has to do with:

- Their anticipatory emotional reaction. For example, many people have severely negative feelings about standing up and giving a speech; others find they tend to clam up in the presence of someone of significantly greater authority. There are nine core communication situations in a working environment (Clutterbuck & Hirst, 2002) and we all have different levels of comfort within each of these situations. The nine form a matrix, with the direction of communication on one axis (to peers, to people more senior, to people more junior) and the size of audience on the other (one-to-one, one-to-group, one-to-many).
- Their level of skill in using the technology. This encompasses both familiarity with instructions and routines and how they adapt other skills and knowledge to enhance their usage. So, for example, someone with graphic arts skills can apply these to the appearance of IT-based communication, or someone with strong literary skills may find it relatively easy to bring colour and tone to the language of email correspondence.

Two substantive academic chapters in recent years have examined e-mentoring from the perspective of Kram's (1985) functions of a mentor (Fagenson-Eland & Lu, 2004; Ensher & Murphy, 2007). While these functions (a sort of amalgam of actions and behaviours) represent only one, narrow, somewhat directive form of mentoring at the detailed level, at the macro level they are generally descriptive of all mentoring. Mentors offer help that is psychosocial (e.g., giving support and encouragement) or career focused.

There does not seem to be any reason why e-mentors (and e-coaches) cannot provide encouragement, career guidance, and role modeling. Distant mentors, such as the arch-epistle-writer St. Paul, have demonstrated this ability through the cumbersome medium of paper-based letters for thousands of years. The difficulties arise mostly with those of Kram's functions that relate most closely to what we now refer to as sponsorship mentoring:

- *Sponsorship* (hands-on help in building a protégé's reputation and career). The e-mentor is much less likely to have direct experience of the protégé's work, so less likely to engage in direct promotion.
- *Giving exposure and visibility, protecting, and giving challenging work assignments.* Again, being at a distance, the e-mentor is less able to influence the protégé's work environment and access to career-enhancing

projects. E-mentoring relationships rarely conflate mentoring with line management responsibilities (although it is possible), and many authorities argue that the two roles are incompatible (Klasen & Clutterbuck, 2001; Cuerrier, 2003) in any form of mentoring.

Other mentor functions include role modeling, counseling (meaning exploring personal issues rather than professional therapy), acceptance and confirmation, and friendship. While e-relationships may not provide opportunities for the mentee to observe specific behaviors of the mentor, this may not be the primary role modeling mechanism in developmental mentoring and developmental coaching, where the emphasis is on the *quality of the learner's thinking*. It can be argued that a text-based discussion gives more opportunities to test the rigour of logic, to explore divergent assumptions and to clarify processes and heuristic approaches.

Counseling has increasingly become an online industry, with specialist courses and guidance (Wright et al., 2008). Valued behaviors by mentors include acting as a sounding board, helping develop coping strategies, and just being there to listen. Sometimes the e-mentor can help without even being involved. We have encountered a number of circumstances where e-mentees and e-coachees have poured out their feelings in an email, left it overnight and found the release of emotion so cathartic that they have then deleted the message in favor of one that is much more focused on what they can do about the situation! While there may be confidentiality concerns about internet conversations, there is no evidence that people are more reluctant to disclose personal information online. Indeed, the remarkable openness of the blogging fraternity demonstrates a surprising lack of concern about personal privacy. In the authors' experience of programs and individual e-coaching and e-mentoring relationships, it sometimes seems easier to open up to someone who is geographically distant and not involved.

Acceptance and confirmation, along with friendship, depend to a considerable extent on the degree of personal rapport and trust that has built up between the mentor and mentee. Fagenson-Eland and Lu (2004) report that their own peer mentoring relationship struggled to achieve these characteristics, because "it was difficult to develop rapport and friendship when online communication was the only avenue available to get to know one another" (p. 153). It is for this reason that many e-mentoring programs arrange for participants to meet face to face at the beginning, to begin to build trust. However, Fagenson-Eland and Lu (2004) conclude that "virtual mentoring can potentially provide most of the functions that face-to-face mentoring provides. If a mentor is willing to help a protégé online, if mentors and protégés are committed to communicating frequently with one another, and if they trust one another enough to share personal as well as professional information and concerns, then effective mentoring can take place" (p. 153).

It should be noted, however, that there is a wide range of approaches to coaching. The more heavily psychology-based approaches rely heavily on a combination of visual, auditory, and language-content clues. Gestalt-based coaching, for example, places importance on the emotional projection that occurs from the coachee and coach. While it isn't impossible to experience the coachee's emotions using non-face-to-face media, it is likely to be much more difficult.

The overview in Tables 1.1 and 1.2 summarizes potential benefits and downsides of face-to-face coaching/mentoring and asynchronous virtual coaching/mentoring.

TABLE 1.1   Face-to-Face Coaching/Mentoring

Benefit	Downside
"Media rich"—more channels of information. Seeing the "whites of someone's eyes" is valuable in getting to know the whole person, rather than the parts that they choose to reveal in text. Building rapport can be very rapid face-to-face, but take much longer by email.	Power dynamics can disrupt open communication. For example, in schools-based mentoring, it is easy for participants to drop into parent–child attitudes and behaviours. Reducing the communication to the message helps promote equality in the exchange. (Especially if the younger person is more competent and knowledgeable about IT processes!) By taking hierarchical and visual cues out of the equation, email makes it easier for people of different age or status to engage in dialogue (Sproull & Kiesler, 1986).
Immediacy of feedback/response in the session itself	Delays caused by practicalities of arranging a meeting
Relative ease of creating "flow"—idea-sparking conversations where each partner feeds off the other	Capturing ideas and thoughts disrupts the flow of conversation. Taking notes attenuates attentiveness drastically. But not having a record makes it difficult to place this conversation into context with those before and after.
Visual cues are important in recognizing when a mentoring relationship is going wrong or running out of steam. Hence face-to-face mentoring is more likely to deal effectively with issues, such as running out of steam. In theory at least, face-to-face developmental relationships should have a higher success rate in terms of longevity and achievement of objectives, but this has not been empirically tested.	Face-to-face meetings can be relatively expensive, especially if they require travel time. *However,* travel time can be an important part of the relationship, providing private space to prepare for the mentoring conversation, and to reflect on it afterwards.

TABLE 1.2   Asynchronous Virtual Coaching/Mentoring

Benefit	Downside
"Just in time" conversations—e-coaches/e-mentees typically communicate as an issue arises and obtain responses within hours or days, instead of waiting for perhaps weeks for a face-to-face meeting.	Participants need to be computer literate and have access to computers. They may also need additional training in how to communicate effectively using email. People with poor general literacy skills, who can hold their own in a verbal conversation, may be discouraged by having to use text as the primary communication medium.
Built-in reflection time—mentors/coaches tend to ask fewer and better questions; mentees tend to consider their responses more deeply	Social networks, which are important in finding suitable mentors informally, tend to operate mostly face-to-face. Fagenson-Eland and Lu (2004) maintain that it is more difficult to "suss out" potential mentors through web networking and that the strength of the social network may be insufficient to persuade potential mentors to accommodate a complete stranger.
Facility to cope with geographical and time separation. E-mentoring and e-coaching allow frequent interaction between partners, who may be on different continents and/or in different time zones. This facility effectively widens the pool of available participants—so mentees, for example, can select the most suitable mentor for them, regardless of location.	In some forms of coaching, feedback from the coach, based on observation of the coachee's work, is an important ingredient in the process. E-coaching typically lacks this opportunity to observe.
Variable length of conversations—because virtual meetings do not need tight timing, they can extend over hours or days without being disruptive; or they can be limited to a relatively short interaction.	Miscommunication. Ensher et al. (2003) suggest that the tonal bareness of email may lead to misunderstandings. Our experience is that this can be worse between speakers with the same first language (e.g., U.S. and U.K. English) than between those with different first languages, because they assume that words have the same meaning. Where language is recognized as a potential problem, people are more likely to query meanings.
Easier group coaching/mentoring. E-partnerships have the potential to create virtual networks, with resultant richer variation in experience, perspective and knowledge sharing.	Slower development of the relationship (Bierema and Merriam, 2002)

Benefit	Downside
Less impact of participants' stereotypes. Say Fagenson-Eland and Lu (2004): "Virtual mentoring...diminishes misunderstandings that arise due to cultural and racial differences. Stereotypes in face-to-face mentoring relationships become invisible in a virtual forum, allowing mentoring to be the focus of the relationship" (p. 155).	The lower visibility of stereotypes may also be a disadvantage in diversity-focused programs. Identifying and working with inaccurate and inappropriate assumptions is a learning fulcrum for the relationship.
A mentee's willingness to provide feedback to a mentor is greater when the relationship is virtual rather than face-to-face (Ang & Cummings, 1994)	Reduced commitment. At least one study suggests that participants find it easier (less embarrassing or emotionally difficult) to disengage from e-mentoring (Whiting & de Janasz, 2004).
Some people are more comfortable communicating online than face-to-face. One study found that introverts achieved more in virtual mentoring groups than in face-to-face groups (Hubschman, 1996). Hamilton and Scandura (2003) also found that people who are shy, unassertive, or simply reluctant to make contact face to face are more likely to do so by email— there is less perceived risk. This may be particularly relevant in cultures where male-female conversations are constrained. Email provides an arm's length relationship compatible with notions of propriety.	Trivialization. The ease of dashing off an email can lead mentees/coachees to make contact on issues that they can easily sort out for themselves. Clear guidelines on when and how to contact the e-mentor/e-coach are essential.

Synchronous (real-time) learning conversations have the benefit that they can stimulate creative thinking, through interactive flow of immediate ideas. They are less effective at stimulating immediate reflection, as there is often a sense of pressure to respond rapidly, if only to let the other person know that you are still there! Most of the other pluses and minuses of asynchronous exchanges also apply.

Telephone coaching/mentoring is also part of the media mix. Used on its own, it has immediacy and the advantage of being able to hear tones and inflections. There is some evidence to support telephone coaching as an effective method for professional skills development (Gatellari et al., 2005), but no comparison exists that we are aware of between telephone-based and face-to-face coaching or mentoring. However, our experience is that it takes a very effective coach or mentor to overcome the inherent difficulties with telephone coaching, which appears to have relatively few of the advantages of either face-to-face or email communication. An interesting comparison comes from studies commissioned by the UK's Learning & Skills Council into e-learning (Hills, 2008). These found that telephone tutoring is the least valued form of support, compared with face-to-face job-related assignments and e-learning resources. Adding video-conferencing or Skype, although it can be rather stiff, helps overcome some of the problems with telephone coaching/mentoring.

Nonetheless, a remarkable proportion of coaching is carried out by telephone (Table 1.3). A recent study by PriceWaterhouseCooper (2007) found that nearly half of 7,000 coaches in 74 countries used primarily telephone coaching and more than half in the United States. A major factor here appears to be commercial: it is easier to pack multiple clients into a day, and there are fewer cancellations.

Experienced telephone coaches rely heavily on intuition and on voice tone. When the client's voice tone changes, they may ask what his or her body position is, as a clue to emotions. It's important to be able to match the client's learning preferences and language preferences, so many telephone coaches place high reliance on processes from NLP. Some clients (and coaches) like to get up and walk about while they think. This is much easier and less disruptive by telephone than face to face.

Finding the Right Mix of Media

Our interviews reveal a wide range of applications, from "pure" asynchronous email-based relationships to some that include every medium available.

Useful questions to ask in deciding which mix of media to use, and when, include:

- What depth of conversation do we want to have? (For deep, transformational conversations, face-to-face conversations will be preferable, followed in order of efficacy by structured email.)
- How frequent and flexible do we want the interactions to be? (Email or telephone tend to give the greatest flexibility and opportunities for ad hoc exchanges.)
- How much time do we want to invest in each session of dialogue? (Above 30 minutes, all media except face-to-face and email become difficult to sustain.)

In practice, an individual relationship is likely to include opportunities for a wide mixture of media, including texting and Skype. (The effective use of media is described in more detail in Chapter 3.)

TABLE 1.3   Telephone-Based Coaching

Benefit	Downside
Immediacy. As with email and other electronic exchanges, a telephone conversation can be "just-in-time."	Problems with silence. Even experienced coaches struggle with maintaining silence while the coachee reflects. Clients typically have much less experience with and capacity to manage silence. The result is that telephone conversations tend to have very little reflective space. (If we can't hear anything, it takes just a few seconds for us to worry that the other person has gone or been cut off.)
Ability to hear tone. It's not just what is said, but how it's said, that provides clues into the client's thinking.	Single sensory input. To some extent, being reliant on sound only can focus the mind in much the same way that people who are visually impaired become more aware of sounds. However, communication is richest when it uses multiple senses and, in particular, a mixture of sound and vision. Very often, coaching/mentoring dialogue requires the use of diagrams, drawings or other visual representations, which are not feasible using telephone alone.
Quality of listening. Just as a blind person becomes more aware of sounds, so experienced telephone coaches and mentors claim to develop greater sensitivity to auditory clues. (Hymer, 1984)	Fatigue. Genuine dialogue by telephone requires the coach or mentor to be at a constant high state of attentiveness, which is not required in any other medium. Being continuously "with" the client can be very tiring, so sessions tend to be relatively short (no more than an hour, at most, versus 90 minutes or two hours for face-to-face sessions).
Rapport-building—the time it takes to build relationships by telephone seems to be mid-way between that for face-to-face and email. However, there is no reliable direct evidence to confirm this.	Satellite delay. While only relevant to really long-distance calls, delays in signal processing can be disruptive, particularly when both parties try to speak over each other.
Conversational flow—though less effective in this respect than face-to-face, telephone conversations can lead to mutual rapid sparking of ideas.	Distractive activity. Especially when on speaker phone, alpha male executives in particular are prone to try to do other tasks, such as reading or downloading emails, when they are not doing the talking.

The Effective E-Mentoring/E-Coaching Conversation

Frequency of communication is a critical factor in maintaining effective virtual developmental relationships and programs (Single & Single, 2005). It's easy for e-partners to become "out of sight, out of mind," so the relationship needs to establish a clear protocol for frequency of contact, even if to say that the coachee/mentee has no issues to discuss at the moment. Most PCs have a reminder function, which alerts users when they have gone beyond the date of expected next contact, and this should be an automatic part of relationship management. (Some centrally administered programs also remind participants if they go beyond, say, 4 weeks without making contact.)

All coaching and mentoring sessions have three sequential phases:

- Preparation before the formal conversation
- The formal conversation itself
- Reflection after the conversation

Both the coach/mentor and the learner need to be fully engaged with each of these phases.

Preparation

A mentee of one of the authors took to the practice of having a pre-conversation. She would place a cushion on a chair in front of her and imagine it was her mentor. As she discussed her issues with this very virtual entity, she took notes, which became the basis for the email she subsequently sent to begin the formal mentoring dialogue. Afterwards, she might have another cushion conversation to develop her thinking around points raised in the "real" conversation.

However they do it, both parties in the virtual developmental relationship should carry out some preparatory reflection. For the coach/ mentor, it may be enough to reflect on how they have helped so far and what they could do better. For the coachee/mentee, it is important to be able to answer these questions:

- What thinking have you done around this issue so far?
- What is your motivation for bringing it into the coaching/mentoring relationship?
- How does this issue link with the "big themes" of your life and work?
- Have you examined this issue from both an intellectual (rational) perspective and an emotional perspective? (i.e., do you know both what you think and what you feel about it?)

In initiating the discussion, the coachee/mentee should consider:

- What do I know and not know about this issue and how it affects me (emotionally and intellectually)?
- What information will help the coach/mentor understand *succinctly* what is going on here?

- What additional information can I provide that will bring this issue to life? (For example, an illustrative anecdote or a metaphor that captures the spirit of the issue.)
- Is this actually one issue, or several mixed together? If the latter, can I separate them, or do I want the coach/mentor to help me do so?

Before sending the email, the coachee/mentee should consider:

- How would I react if I received this?
- Is there a clear structure to the message? (It can sometimes be good practice to indicate what structure you are working toward.)
- What kind of response do I want from the coach/mentor? Have I made this clear to him or her?

A common reaction is that the message is too whinging, or that the issue seems too trivial when put on paper. Don't be put off. Let the message marinate for a while, and then when you return to it, make adjustments. If you still have worries about the tone or subject, express these in an initial paragraph. Doing so provides valuable information to the coach/mentor about your state of mind, the depth of your concerns, and your mood. This emotional content is also important in building up the level of relationship rapport.

When the mentor receives the email, it is usually best not to reply immediately. Again, let the issue marinate, although you may wish to respond to simple information requests, such as for a contact address, fairly quickly. Consider:

- What is the learner saying here?
- What is he not saying? What is missing from the picture? What more information do I need to understand what is going on?
- Is she being honest with herself? With me?
- Are there contradictions and/or non-sequiturs in his account?
- How relevant is my own experience? How can I use that experience without telling her what to do?
- What questions would make him think more deeply about the issue?
- What is she looking for from me? Is that expectation appropriate? If I offer her a different response, how will I help her understand why it is better?
- What other sources of information or help can I put him in touch with?

As with face-to-face coaching/mentoring, there may also on rare occasions be issues that touch the boundaries of safe and appropriate intervention—for example, if the learner appears to be severely emotionally disturbed. Making judgments about when to refer a coachee/mentee on to professional psychotherapy is a lot easier face-to-face. Some signs to watch for, however, are when:

- Messages are incoherent, rambling and contradictory,; or irrationally assigning blame to everyone else and either none or reluctantly to the learner.
- The language is too precise and controlled, giving the impression that the learner is trying to present a "public" image and preventing you from seeing their private image.
- The learner refuses to discuss emotional issues.
- The tone of the message suggests that the learner is shouting at you.

Whenever the e-coach/e-mentor has suspicions of psychological health problems, he or she should contact either the program coordinator or a professional coaching/mentoring supervisor. All virtual development programs should have qualified supervision as an online ad hoc resource. Good practice involves having at least two supervisors available, with different strengths and backgrounds (for example, one with extensive coaching/mentoring expertise, one with psychotherapy qualifications).

The ability to analyze email text provides an opportunity for the mentor to explore language and content more deeply. What words and phrases recur frequently, not just in the current message, but in previous messages? Are there patterns of association between these recurrences and particular situations, individuals or emotional states? Once you start to look for these links, it's amazing how obvious they sometimes become.

Always read the message at least three times before sending your reply:

- Once to get the drift
- Once to craft a response
- Once more to assess your response against the initial message. (Will this reply really be helpful? How would I feel if I received this? What have I held back from saying and why? What do I want the learner to do as a result of receiving this?)

Reading the message several times also helps identify ambiguous words or phrases—or simply expressions, where you suspect the learner may be making assumptions different to your own. It is important to request clarification: "What precisely do you mean by this word?"

The virtual coaching/mentoring session may have several exchanges or just one. Between each exchange, there should be a period of reflection, and a much longer period of reflection between sessions. In this longer reflection, the e-learner should consider:

- What have I learned from this conversation?
- What does this change for me?
- What do I want to do (differently)?
- How well did I help the e-coach/e-mentor help me? What could I do better next time? (For example, how can I make my intent clearer?)
- What do I want to bring to our next conversation?
- What can I feed back to the e-coach/e-mentor about what he or she did that will a) show my appreciation and b) help him or her be effective next session?

Similarly, e-coach/e-mentors need to reflect on how they managed their part of the conversation. For example:

- Did I truly understand the issue presented?
- For whose benefit did I ask each question?
- How helpful were the questions I asked? (Did they provoke thoughtful responses?)
- If I offered advice, how appropriate was it?
- Did I take enough time and mental space to reflect before replying?
- How do I feel generally about this e-conversation?
- What could I usefully take to a supervisor for discussion?

In all coaching and mentoring, it is good practice to review the relationship on a regular basis. Doing this online once again allows for more reflection time. Useful questions to ask of each other are:

- What did each of us do that was most/least helpful from a practical perspective?
- What did each of us do that most/least stimulated thinking?
- What should each of us do more of?
- What additional support, if any, would help us make more of this learning alliance?
- Have we achieved an appropriate level of trust and mutual respect?
- Were there occasions when one or both of us felt that the communication was missing something? Or maybe we were on a different wavelength? What did we do about it and what could we do in future?
- How clear are we about what the relationship is for and where it is going?

The Skills of Virtual Coaching and Mentoring

For anyone who has read so far, it won't be a surprise that the skills needed for e-coaching and e-mentoring are much the same as for their face-to-face counterparts, with a few more added. If anything, the level of skill required in the virtual environment is more equal between coach and coachee/mentor and mentee than in face-to-face programs.

Having a strong interest in developing other people is critical for coaches/mentors. Having a strong commitment to one's own development is equally important, but this time for both parties. One of the most useful things a learning partnership can do in building rapport, mutual respect, and a sense of shared purpose is to exchange their learning ambitions, both for the relationship and more generally, in an early email exchange.

In very much the same way, having a sense of one's own values is important in rapport-building. By disclosing their own values and helping the coachee/mentee discover and explore their own, the mentor or coach establishes a tone to the relationship that goes way beyond the transactional.

Communication skills—both general and computer-specific—are also essential. These include effective use of questions, metaphor, and stories; giving feedback; and structuring information logically and clearly. Listening skills, so essential in face-to-face coaching/mentoring, are replaced by a wider attentiveness—an ability to see meaning beyond the written word. Many of the most successful e-coaches and e-mentors we have spoken with find that they make much more frequent and effective use of intuition. They also pay considerable attention in each relationship to how they build and demonstrate empathy.

Computer skills can be fairly basic, as long as they are sufficient for the role. Many senior people are declared or closet technophobes—it takes courage to ask for help in making the most effective use of Skype or learning how to produce simple drawings on the screen.

Relationship management—for example, knowing when to cajole and encourage, when to back off—is subtly different in e-coaching and e-mentoring, compared with face-to-face. In face-to-face developmental mentoring, the mentee is expected to take gradual charge of relationship management. In sponsorship mentoring, the responsibility may rest more substantively with the mentor, but the same aspiration applies. In e-coaching and e-mentoring, however, both parties typically need to be equally proactive right from the start. Discussion about how to make the relationship work therefore needs to assume a much greater role in initial training and negotiations within the dyads.

A final skill required by virtual coaches and mentors is patience. Asking a thought-provoking, massively difficult question in a face-to-face developmental relationship may require at most a few minutes' silence as the learner thinks things through. In e-mentoring and e-coaching, either party may take days (or even longer) to respond. This is not necessarily because they are not taking the message seriously—more likely, it is because they are giving the serious thinking time it requires!

Emotional Content in E-Mentoring and E-Coaching

Communicating emotion is important, because it provides context to the content of a message or conversation. Context may either relate to:

- The emotional tone (positive, negative or neutral)
- The type of emotion intended (e.g., anger or pleasure)
- The intensity of the emotion (mild to strong)

Without emotional content, it is more difficult to identify accurately the other person's intent. It also poses problems with building trust and rapport. Email is perceived as having fewer emotional cues and less feedback on how emotional intent is received than face-to-face communication. The delay in response may also interfere with the accuracy of emotional interpretation.

Kristin Byron's (2008) analysis of the literature on email communication identifies a range of issues that have relevance to e-coaching and e-mentoring. One of the issues that is often overlooked is difference in gender responses to media. A study comparing face-to-face and email conversations on the basis of persuasiveness found significant gender differences (Guadagno & Cialdini, 2002). The researchers found that women form stronger communal bonds in face-to-face conversation, and that these bonds enhance their ability to influence each other. However, email disrupts the exchange of social cues they use to build bonds, making it less easy to persuade and be persuaded. Men, by contrast, react in groups by attempting to establish their individual competence, and hence are more confrontational, so they form fewer and/or weaker bonds. Subsequent conversations allow greater room for persuasion, if they are held electronically rather than face to face, because there is less stimulus to compete.

What We Know about Emotions and Emails

- Email senders do communicate emotion, intentionally or not (Walther & D'Addario, 2001), but it is likely to be ignored or misinterpreted.
- People find it easier to identify cues to negative emotions in emails than those to positive emotions (Byron & Baldridge, 2005).

Byron proposes that email recipients interpret messages intended to convey positive emotions more neutrally than the sender intended. In part, this is because email messages produce less physical arousal than face-to-face (Kock, 2005). Similarly, neutral or mildly negative messages tend to be interpreted more intensively negatively than intended. There appears to be a gender effect here, too. Men use fewer emotional cues in emails; their emails tend to be emotionally neutral. Emotional expression is also affected by age (older people tend to be less emotionally expressive overall in their communication); and by relative status (employees may have lower expectations of positive emotions in emails from people at more senior levels in an organisation).

Implications for E-Coaching and E-Mentoring

The e-coach/e-mentor needs to be aware of this aspect of e-communication and compensate accordingly. Using emoticons can help, but they may seem artificial and distracting. More sensitive approaches include:

- Using emotional language (e.g., "I was really pleased" or "I was deeply concerned") more frequently
- Asking questions to assess the emotional depth and tone of statements by the coachee/mentee (e.g., "How deeply do you feel this?" or "Do you think this is just a passing emotion, or one that will affect you for some time?")
- Giving emotional feedback ("I'm interpreting your comments as being really angry—is this so?")
- Emotional contracting—agreeing as part of the relationship set-up that you will both address feelings openly as part of the e-conversation

Of course, it is important to recognize that the emotional flatness of email can also be a strength. There are times when it is valuable to be able to step back from the emotional into a rational perspective. Also, it can sometimes help the coachee/mentee to encourage him or her to write on the computer all the emotion they feel about an issue—getting it off their chest—and then erase the file. Doing so often seems to help in erasing or at least blunting the negative emotions and the coachee/mentee is more able to discuss these less intense emotions from a more balanced perspective.

SUMMARY

In this chapter we have explored some of the reasons why e-coaching and e-mentoring have grown in popularity and tried to define what we mean by the terms. We have analyzed some of the strengths and weaknesses in comparison with more traditional approaches to coaching and mentoring, and we have reviewed some good practice in managing the virtual developmental relationship. In the next chapter, we discuss how to design and implement effective, sustainable e-coaching and e-mentoring programs.

NOTES

1. U.S. academic articles typically use the term *protégé* (meaning someone who is protected). European and many practitioner sources tend to use the term *mentee* (defined as one who is helped to think). The two terms reflect the difference between sponsorship mentoring (where the power and authority of the mentor is a significant factor) and developmental mentoring (where it is not).
2. Actually a composite of more than one person

REFERENCES

Ang, S., & Cummings, L.L. (1994). Panel analysis of feedback-seeking patterns in face-to-face, computer-mediated and computer-generated communication environments. *Perceptual and Motor Skills, 79*(1), 67.

Bierema, L.L., & Merriam, S.B. (2002). E-mentoring: Using computer mediated communication to enhance the mentoring process. *Innovative Higher Education, 26*(3), 211–227.

Byron, K. (2008). Carrying too heavy a load? The communication and miscommunication of emotion by email. *Academy of Management Review, 33*(2), 309–327.

Byron, K., & Baldridge, D. (2005). Toward a model of nonverbal cues and emotion in email. *Academy of Management Best Papers Proceedings,* OCIS B1-B6.

Clutterbuck, D. (2007). A longitudinal study of the effectiveness of developmental mentoring. Ph.D. thesis. King's College, London.

Clutterbuck, D., & Megginson, D. (1999). *Mentoring Executives and Directors.* Oxford: Butterworth Heinemann.

Clutterbuck, D., & Hirst, S. (2002). *Talking business: Making communication work.* Oxford: Butterworth-Heinemann.

Cuerrier, C. (2003). *Starting point for mentoring.* Quebec: Fondation de l'entrepreneurship.

Emelo, R. (2008). Work-based mentoring's impact on retention and productivity. Triple Creek Associates, www.3creek.com

Ensher, E.A., & Murphy, S.E. (2007). E-mentoring. In Ragins, B.R. & Kram, K.E. (Eds.), *The Handbook of Mentoring at Work: Theory, Research and Practice* (pp. 299–322). Los Angeles: Sage.

Fagenson-Eland, E. & Lu, R.Y. (2004). Virtual Mentoring. In Clutterbuck, D. & Lane, G. (Eds) *The Situational Mentor* (pp. 148–159). Aldershot, UK: Gower.

Ferrar, P. (2006). The paradox of manager as coach: Does being a manager inhibit effective coaching ? (Unpublished Masters dissertation, Oxford Brookes University.)

Gattellari, M., Donnelly, N., Taylor, N., Meerkin, M., Hirst, G., & Ward, J. (2005). Does 'peer coaching' increase GP capacity to promote informed decision-making about PSA screening? A cluster randomised trial. *Family Practice, 22*(3), 253–265.

Guadagno, R.,& Cialdini, R. (2002). Online persuasion: An examination of differences in computer-mediated interpersonal influence. *Group Dynamics Theory, Research & Practice, 6*, 38–51.

Hamilton, B.A., & Scandura, T.A. (2003). Implications for Organizational Learning and Development in a Wired World. *Organizational Dynamics, 31*(4), 388–402.

Harrington, A. (1999). E-mentoring: The advantages and disadvantages of using email to support distant mentoring. www.coachingnetwork.org.uk/Resource-Centre/Articles/viewarticle,asp?artId=63

Hills, H. (2008). Achieving maturity in e-learning. *Training Journal, January*, 37–41.

Hubschman, B.G. (1996). The effect of mentoring electronic mail on student achievement and attitudes in a graduate course in education research. *Dissertation Abstracts International* 57-08A p3417

Hymer, S.M. (1984). The telephone session and the telephone between sessions. *Psychotherapy in Private Practice, 2*(3), 51–65.

Klasen, N., & Clutterbuck, D. (2001). *Implementing mentoring schemes.* Oxford: Butterworth Heinemann.

Kock, N. (2005). Media richness or media naturalness? The evolution of our biological communication apparatus and its influence on our behaviour toward e-communication tools. *IEEE Transactions on Professional Communication, 48*, 117–130.

Kram, K. (1985). *Mentoring at work: Developmental relationships in organisational life.* Glenview, IL: Scott, Foresman.

Markus, M.L. (1994). Electronic mail as the medium of managerial choice. *Organization Science, 5*(4), 502–727.

Megginson, D. (2006). An own-goal for coaches. Paper to 13th European Mentoring and Coaching Council Conference, Cologne.

Perren, L. (2003). The role of e-mentoring in entrepreneurial education and support: A meta-review of academic literature, *Education + Training, 45*(8/9), 517–525.

PricewaterhouseCooper. (2007). ICF Global Coaching Study. www.coachfederation.org

Reid, F.J.M., Mallinek, V., Stott, C.J.T. & Evans, JStBT. (1996). The messaging threshold in computer-mediated communication. *Ergonomics, 39*(8), 1017–1037.

Sarbaugh-Thompson, M., & Feldman, M.S. (1998). Electronic mail and organizational communication: Does saying "hi" really matter? *Organizational Science, 9*, 689–698.

Single, P.B., & Muller, C.B. (2001). When email and mentoring unite: the implementation of a nationwide electronic mentoring programme. In Stromei,

L.K. (Ed.), *Creating Mentoring and Coaching Programmes* (pp. 107–122). Alexandria, VA: American Society for Training and Development.

Single, P. B., & Single, R. M. (2005). Mentoring and the technology revolution: How face-to-face mentoring sets the stage for e-mentoring. In F. K. Kochan & J. T. Pascarelli (Eds.), *Creating successful telementoring programs* (pp. 7–27). Greenwich, CT: Information Age.

Sproull, L. & Kiesler, S. (1986). Reducing social context cues: Electronic mail in organizational communication. *Management Science, 32*(11), 1492–1512.

Stokes. P., Garrett-Harris, R., & Hunt, K. (2003, November). *An evaluation of electronic mentoring (e-mentoring).* European Mentoring and Coaching Council Conference.

Walther, J.B., & D'Addario, K.P. (2001). The impacts of emoticons on message interpretation in computer-mediated communication. *Social Science Computer Review, 19*, 324–347.

Whiting, V & de Janasz, S.C. (2004). Mentoring in the 21st century: Using the internet to build skills and networks. *Journal of Management Education, 28*, 275–293.

Wright, J., Gooder, D., & Lang, S. (2008). A mixed blessing? Using email for counseling psychotherapy at a distance. *Counselling, Psychotherapy, and Health, 4*(1), Counselling in the Asia Pacific Rim: A coming Together of Neighbours Special Issue, 54–63.

CHAPTER 2

MAKING THE VIRTUAL DEVELOPMENTAL PROGRAM WORK

David Clutterbuck

As we have already seen, e-coaching and e-mentoring programs come in many forms, so there is no overall set of standards that covers all the major permutations. However, there is an accumulating body of practical experience from effective and less effective programs from which we can extrapolate broad guidelines as to good practice. The advice we give here is primarily aimed at people responsible for creating and maintaining programs within an organization or on behalf of a defined client group in society.

PREPARATION

The most common problem we encounter in any mentoring or coaching program is lack of preparation and forethought—working out exactly how things will work, ensuring sufficient appropriate resources, and creating contingency plans. In a virtual mentoring/coaching environment, particularly one that builds or adopts a technology platform to support the process, preparation failures can be much more difficult to overcome.

Virtual Coach, Virtual Mentor, pages 31–51

It's important to do some very practical homework first. Key questions include:

- What are the critical transitions the beneficiaries of the program go through, over what period? For example, prisoners leaving jail may need one form of support to prepare for their release, another a few weeks after release when the reality of their new situation hits them, and a different approach yet again when they have found a job and need to establish personal management processes to ensure they don't regress into old habits.
- Who are the beneficiaries most likely to listen to, respect, and accept as role models?
- Why is virtual coaching/mentoring an appropriate solution in this context? Will it work best as a "pure" on-line program, or should there be elements of face-to-face in it?
- Can and should we train and support coaches/mentors on-line? How cost-effective is this?
- What level of technical competence will the participants have?
- How complex a technological base will we need? Does the program require a customized IT support platform, to mediate between participants and monitor, or can we simply match people and swap e-mail addresses?
- What experience or perceptions of coaching/mentoring do the potential participants have, if any? (Previous negative experiences create an extra barrier to overcome.)
- What else could get in the way? (Some time ago, one of the authors proposed an e-mentoring program between female professionals in the U.K. and in Kabul, Afghanistan. Unfortunately, very few of the target audience in Kabul spoke English and Afghani speakers in the UK were mostly refugees struggling to find work in their professions.)
- What support do we have for the program? Where are the champions going to come from, and will they be able to provide the resources and influence the project needs?
- What other programs have addressed this target group elsewhere? What challenges did they face? Where did they succeed and where did they fail?
- How will we ensure that the program is sustainable? The landscape of coaching and mentoring is littered with the dried husks of programs that flourished for a while, then withered when the funding stream ran out.
- How will we know this program has been successful? What measures should we apply and why?

Questions such as these should ideally be addressed in the initial funding request. In practice, there is often insufficient application time—especially in a corporate environment—to gather all this information. However, the recurrent message from participants on Clutterbuck Associates' public Programme Coordinator courses is that every hour spent on preparation saves many times the time and money in project management and troubleshooting; and increases the chances of delivering the program objectives.

A detailed project plan is essential. The headings that follow cover the most critical elements of the plan.

PURPOSE

What is the point or intention of this program? Who precisely is it aimed at and what specific changes do you want to bring about in them over what period of time? Some of the biggest disasters we have seen have occurred in programs where the target group wasn't clearly defined, or didn't see themselves as having similar issues, and where the kind of changes intended required two or three times the allotted relationship length and/or twice the intensity of interaction. For example, six months of monthly meetings is likely to be inadequate for severely alienated young persons at risk.

Based on the preparation work already done (hopefully) it should be possible to define very clearly:

- How an e-coaching/e-mentoring program will help this specific group of beneficiaries
- What the wider benefits will be (e.g., to the company, or to society)

Recent research (Clutterbuck, 2007) suggests that, for mentoring at least, understanding the wider organizational or societal objectives underpinning the program influences the quality of participants' experience. Why this association occurs is open to debate, but likely explanations include:

- People like to feel the relationship is supported, without suspicion about motives
- Learning relationships require a shared sense of purpose, if they are to fulfil their potential.

ROLES: COORDINATOR, STEERING GROUP

The coordinator has a pivotal role in the project (Clutterbuck, 2006). He or she promotes the program to participants and corporate sponsors; protects it from the destructive forces of sudden economies and changes of management fashion; and maintains the quality of the program, often on limited budgets.

Some key issues to consider:

- What on-line access does the coordinator need to maintain a close awareness of the health of the program and individual relationships?
- Who else (e.g., members of the steering group) requires access at a macro-level?

RESOURCES: FINANCE, IP, TECHNOLOGY

The key financial issues are:

- How much?
- Where from?
- What are the accountabilities? (What will you have to report back on and to whom?)

The cost will depend on a number of factors, but each element of the program considered here will have a cost. Some of the specific issues include:

- Recruitment. Within a corporate environment or a professional body, recruitment costs can be relatively low—a matter of sending out e-mails and holding a few information evenings or lunchtime briefings. With public programs, however, the costs of attracting and vetting applicants can rise very steeply. For example, finding successful women entrepreneurs to mentor younger businesswomen might cost in excess of $8,000 per head. Check with other, similar programs to establish actual costs.
- IT costs. Even if you do not have to provide participants with access to PCs, the technology costs can be very high. Packaged IT solutions usually need customization and tend to come with bells and whistles you do not need—but they are typically cheaper than building your own.
- Staff costs. It takes far more time to administer an effective coaching or mentoring program than people expect. As a rough rule of thumb, a virtual mentoring program will need an average of a day a week of administrative time, spread over a year, for every 20–25 pairs. The same would apply in most cases to coaching programs with a strong personal developmental content, but technical skills coaching programs typically require much less administrative input.

(Most packaged IT solutions come into their own particularly for skills and knowledge transfer.)

- Training costs. Whatever you do, don't skimp on the training! The quality of the relationships depends heavily on participants' confidence and competence in fulfilling their roles. If you bring in external trainers, it is best to use people who specialize in this area and who already work on programs accredited by the EMCC, ISMPE, ICF or similar bodies. Specialists may cost 10–20% more, but they will be able to add far more value, for example, in the quality of advice they are able to give. If you decide to use your own trainers, consider putting them through a train-the-trainer program from a specialist provider. The relatively small cost of attending a program coordinator's accreditation course is similarly recommended as good value for the money.
- Participant costs. Will you cover expenses? In virtual coaching/ mentoring, these are typically very small (unlike face-to-face mentoring, where they can be a major part of the overall project cost). It is prudent to set aside a budget for expenses and to monitor this in the pilot phase.

Funding is the Achilles heel of many—perhaps most—virtual coaching and mentoring programs. Key questions include:

- Who needs to sign off on the funding?
- How long between application and knowing whether it will be made available? Between then and receiving the funds?
- How secure is it? (Could it be withdrawn at short notice?)
- Will we have to find some of the funding from other resources? (The concept of matched funding is common in funding from the European Union, for example.)
- Will we have to spend the funds by a specific date?
- If we receive only part of the funding, what will we cut from our plans? Experience suggests that the key elements to cut, in order of importance are: numbers (always seek quality over quantity), technology (make it simpler), and staff (make more use of volunteers and steering groups—but don't think you can succeed without at least a part-time qualified program coordinator).

The level of financial accountability required from the funder is usually higher in the public sector than the private sector, where a high level of transparency is usually required. It is important, therefore, to accompany any application for funds with a clear and precise description of how and when the impact of the program will be measured. The more this can be tied to the objectives of the funder, the better. In private sector companies, too, this upfront clarity and accountability makes sponsors more comfortable with the project as a whole. Carrying out the evaluation on-line fits neatly with the ethos of a virtual program.

Intellectual property issues didn't appear in any of the guidance books and on-line advice resources we accessed in our research. But ignoring them can lead to serious problems—not least being sued for breach of copyright! Coaching and mentoring are evolving rapidly, and what constituted good practice and an appropriate understanding of the role five years ago will probably not be recommended now. The latest thinking will almost always be covered by copyright. Key questions to consider include:

- What information, exercises, and diagrams do we want to include in our training materials?
- Do we want to provide a support resource, with relevant articles about virtual coaching/mentoring?
- Who owns the copyright on those materials?

The simplest solution is often to buy in these resources from an experienced provider. This has the benefit that most of the bugs will have been taken out already. Issues to consider:

- How up-to-date is this resource? (Is it continuously updated from research and other sources?)
- Is it accredited with one of the primary bodies in the field, or a university with a dedicated coaching and mentoring faculty?
- How easy is it to use?
- What are the hidden costs, if any?
- Can you buy the resource outright, or do you have to buy blocks of subscriptions?
- Do you have to hold these materials on your own server, or are they maintained centrally by the copyright holder? (If the latter, they take on all the effort of keeping them up to date. You may also be able to have your own portal, so it is not obvious to any other user that you are not hosting the website.)

If you decide to create your own resources, remember that you have to gain permission from the copyright holder to use anything but short quotes that you extract from elsewhere. (Five lines is a standard guideline.) Redrawing diagrams and charts with minor changes is not enough to overcome copyright requirements. The first port of call in seeking copyright permission is normally the publisher (not the author). Some publishers—for example, *Harvard Business Review*—charge fees, which community program managers may find beyond their budget. A cheaper option is to access authors' own websites for articles and other materials and to ask permission for these. As long as the borrowing is relatively insubstantial, most authors are happy to allow you to reprint their text without charge. Whenever seeking permission to reproduce, state where you want to use the materials, who and how large the intended audience will be, the objective of the program/ resource, and the exact page reference, along with a copy of the relevant section of the document you have prepared (ideally, the pages using the quoted text or illustration, plus one or two pages on either side).

The technology we will discuss in greater detail in the next chapter. However, critical questions to address at an early stage include:

- Does everyone have access to compatible IT equipment? Will this be their own or a shared resource? (Shared resources can potentially create confidentiality problems in the absence of strict access protocols.)
- Will they have to store some programs or materials on their own computers?
- How frequently do you expect participant pairs to make contact, and for how long?
- Do you envision some group discussions between coachees/mentees about common issues? If so, do you want these to be moderated?
- Will program management software be integrated with participant access software? If so, where will "Chinese walls" be required and how robust must they be?
- What software can you buy off the shelf and what will need to be customized? How do the various offerings compare on price (initial and on-going), flexibility, and support systems? Of the facilities offered, what do you really need and what are simply nice to have?
- If you need customized software, what will it cost and how long will the customization take?
- Is there a trade-off between functionality and ease of use? Most people don't use a fraction of the functions on their PC. The more complex the system, the more likely people are to become frustrated by it; and the more time the program staff will have to spend dealing with queries.
- Who will be providing technical support, and what will this realistically cost? (Whatever figure you think of initially, double or treble it!)
- How will you update the system, and who will do it?
- How long a period will you allow for embedding the system? How will you test it before the program launches, and what will this cost?

COMMUNICATION/PUBLICITY

Professional communication expertise is always helpful in virtual coaching/ mentoring. It is important to research how receptive the target audiences will be to the program messages and to adapt communication accordingly. It also helps to use as wide a range of media as practical, from print to texting, and to encourage peer groups to pass on invitations to find out more about the program. In some programs, up to 80% of volunteers have come from this kind of pass-along.

The communication plan needs to be detailed and to cover at least the first year of the program. Each of the key audiences—participants, third party stakeholders such as mentees' line managers, sponsors and so on— will need customized communication as the program progresses. The more information you have from monitoring and evaluation, the easier it will be to keep all these people engaged with the program.

SELECTING PARTICIPANTS

The program definition should provide a description of the people you want to attract for the coach/mentor and coachee/mentee roles. Indeed, there should be a detailed job description for everyone involved in the program, including the steering group, the coordinator, and the sponsors, and these should be published where potential participants can easily access them.

Among the key issues to consider with regard to selection are:

- How many coaches per coach/mentees per mentor? Most programs with a strong behavioral and/or personal support emphasis tend to look for single pairings, or at most two mentees/coachees for every mentor or coach. Programs aimed at skills and knowledge transfer can usually manage larger numbers per virtual coach. You will need to research how much time will be needed and how much the coaches have available. Using e-mail has another benefit for multiple coaching—the coach can hold on-line tuition sessions on themes that are of relevance and interest to his or her coachee group as a whole, and stimulate peer learning between the coachees in a way that is much more difficult using face-to-face meetings.
- In some circumstances, you may wish to allow (or even encourage) coachees/mentees to acquire more than one coach/mentor, running in parallel. For example, they may have a long-term need to develop leadership thinking patterns, but a short-term need to acquire negotiating skills. What rules or guidelines will you introduce to ensure that this doesn't get out of hand?

- Relevance of eligibility requirements. For example, one company insisted that mentees should be nominated by their line managers. When they removed this restriction in favor of self-nomination, the quality of candidates improved immediately.
- Equal opportunities and diversity considerations. Some programs are specifically aimed at promoting diversity. Those that aren't need to ensure that they:
  - Do not in any way disadvantage any group on the basis of race, gender, culture, religion, disability, and so on
  - Adhere to the letter and spirit of both legal requirements and any relevant codes of practice.

It is a wise precaution to take professional advice on both the selection process and the language used in program communications—especially those concerned with publicizing the program and attracting participants.

- Involve the participant groups in the recruitment process—peer pressure is a powerful positive force in this respect. Again, using the technology available makes practical sense. For example, although there are confidentiality issues, creating a "Mentor's Blog" can create links to all sorts of unexpected volunteer resources.
- Background checks on volunteers. For some programs (e.g., those involving young or vulnerable people), many countries have legal requirements for police checks. It often happens that relationships are delayed in starting because these checks take longer than expected, so build this time period into your program plan. In what circumstances will a criminal record be a reason for denying someone a place in the program?
- What other characteristics will rule people out or in? Work with stakeholders and others to develop a description of the ideal and worst case candidates, which you can publish. Many unsuitable candidates will drop out at this point, but beware that you don't put off good candidates who are overly self-critical or self-deprecating!
- What processes will we use to assess suitability of participants? Options include:
  - Assessment centers, although these are difficult to do on-line and are expensive to design
  - Telephone interview (in which case, what questions will you ask and why? What will these tell you about their suitability?)
  - Assessment by a psychologist
  - Using face-to-face training to observe and assess

In general, this is the point in a virtual learning relationship when some face-to-face interaction is essential. If this isn't possible—for example, if candidates are widely scattered geographically—then an alternative is to arrange for them to e-coach/e-mentor a member of the program team, before (or preferably after) initial training.

- How will we manage candidates we turn down? In an e-community, news travels fast. So the rejection process has to be handled with sensitivity and honesty. Even if the rejection itself is handled by e-mail, it should be followed up rapidly by a telephone conversation, ideally led by a coaching/mentoring supervisor. The rejected candidates need to know:
    - The positives about their approach/attitude, etc.
    - The specific reasons why they were not accepted onto the programme
    - What they can do to address those issues and where they can find appropriate help
    - That the program team is grateful for the interest they have shown

  Don't simply add them to the mentor pool but never use them. Do keep them informed about the project's progress. Where the process has been handled well, rejected candidates are likely to recommend participation to colleagues whom they see as more suitable than themselves, and to embark on self-development based on the feedback they have received.

- How similar do the coachee/mentee groups and the coach/mentor groups have to be? In some environments, similarity of experience, culture, or background is of high importance, in order to create and sustain rapport; in others, difference is the primary driver of learning. In general, people from similar racial or cultural backgrounds develop relationships that provide greater emotional (psychosocial) support; pairings from different backgrounds tend to place more emphasis on "practical" goals and outcomes.[1] Networking in the former situation has a higher social content; in the latter, it is more associated with access to power and influence. The research that draws these conclusions is, however, based firmly in face-to-face relationships. It could be argued that virtual relationships will be less influenced by similarity or difference. Our own view is that the ability to empathise with the learner's current experiences is critical to both face-to-face and virtual relationships and that this must be taken into account in making the matching decision.

- How much effort will people be prepared to invest in the application process? It pays to research this ahead of time and to be sure to tell people accurately how long filling in the questionnaire is likely to take, so they can set aside enough time. If they give up half-way, they probably won't come back. In specifying your software, you may

wish to insist that there is a "pause and continue" option, so users don't have to go back to the beginning when they log on again.
- How will we take poor coaches/mentors out of the pool?

THE MATCHING PROCESS

Key questions here include:

- How automated do we want the matching process to be? Choices range from a centrally managed process, where simple software allows the program coordinator to identify potential pairings and inform them of the match, to fully automated systems, where learners search a database of potential coaches/mentors and select one or more that they like. The advantage of self-selection is that it saves administrative time; the disadvantage is that learners may select on the wrong basis— for example, they may be drawn to a more senior person on the assumption that he or she will assist directly with career advancement; or they may dismiss the more senior person as likely to be too busy.
- What matching criteria should we include? Again, initial research is invaluable in determining which criteria are most relevant. The most common are demographics, education, and work experience; however, recent learning, current learning objectives, previous experience as a coach/mentor, and out-of-work interests can also be highly relevant. Some matching systems include psychometric data—for example, on personality, learning styles, and communication styles—and in a few cases, how the coach/mentor rates his or her coaching/mentoring ability on a self-assessment questionnaire. All of this data can be built into the matching software, but beware of making the process too complex and time consuming.
- What is our process for re-matching when relationships don't work out?

PROGRAMME MANAGEMENT

Confidentiality is an issue in all mentoring programmes and in coaching out of the reporting line. The general rule is that everything said between mentor and mentee is private, unless otherwise prescribed. Obvious boundaries to confidentiality include and revelation of:

- Criminal activity
- Situations where real or potential harm may be caused to the coachee/ mentee or to other people (e.g., a doctor with an alcoholism problem)

- Behavior that would bring the sponsoring organization into disrepute (e.g., sexual harassment, bullying)
- Serious issues of other forms of professional misconduct (e.g., an accountant fiddling his or her expenses).

However, virtual coaching and mentoring have additional issues with regard to confidentiality. Firstly, e-mail is not secure. Many companies have monitoring software, to ensure that e-mail is not being misused. Participants therefore need to be aware that their conversations may be monitored—and that some confidences may best be communicated by telephone or face to face. Programs aimed at under-16s and vulnerable adults, or at people in jail, also need to be monitored, and some programs have automated checks on this. The Staffordshire University project has a strong statement on confidentiality in its guidance notes:

> In order to provide support for mentors, mentees and the team, information may be shared in confidence with members of staff on the e-mentoring team. Breeches of confidentiality within this team structure may result in the expulsion of a mentor or mentee from the scheme, or for retraining on the part of the mentor (at discretion of staff on the e-mentoring team). E-mail messages between mentors and mentees may, at the discretion of the team and in total confidentiality, be shared with a member of the e-mentoring staff team, if there is a concern about a student's behaviour or well-being and in order to evaluate the effectiveness of the scheme and assist in its future development.[2]

A somewhat neglected aspect of e-mentoring programs, relating to confidentiality, is the use of professional supervision to help develop the competence of coaches/mentors and to provide additional protection to coachees/mentees. The common responses of program managers, when this is pointed out, are that (a) there aren't enough funds for this, and (b) the coaches/mentors are not professional—and hence aren't bound by the rules of organizations such as the European Mentoring & Coaching Council or the International Coach Federation. However, supervision has an important role in maintaining quality within the program and in determining when clients have needs requiring other forms of professional help (for example, therapeutic). The supervisor becomes the ultimate confidante. Where the program does include supervision, the program design needs to address the questions:

- In what circumstances does the supervisor bring issues from his or her conversations with coaches for discussion with the programme coordinator or steering group, and how?
- Is group supervision or one-to-one supervision most appropriate? (Cost usually draws programs towards group supervision, supplemented by individual supervision as needed.)

- Should the program coordinator be informed automatically when a mentor/coach seeks an individual supervision session?
- Can some or all (individual) supervision be carried out by e-mail or telephone?

Effective program management also requires a number of written policies that clarify:

- Confidentiality, as discussed above
- Risk assessment and risk management
- Diversity (see below)
- Health & safety, if appropriate
- How participants can report concerns. If the process is unclear, or participants lack confidence in it, relatively minor relationship problems can grow to a point of seriousness, for example.

In addition, there needs to be a published Code of Conduct, which covers acceptable and unacceptable behavior overall. This may include frequency of e-conversations, content of conversations, and what is and isn't appropriate for the coach/mentor to do on the learner's behalf. The online conversation requires a number of elements in the code of conduct, which may be different to face-to-face mentoring, including:

- When and how frequently it is appropriate for the learner to contact the coach/mentor
- The length of e-mails and the volume of attachments
- Protocols for sharing data within the relationship and with third parties
- In young person coaching/mentoring, ensuring their access to the internet is safe

Often called *netiquette*, these issues are discussed in more detail in Chapter 3.

PROCESS MANAGEMENT

A plus (or a minus, depending on how you look at it) of virtual correspondence is that inappropriate language is obvious to any monitoring. For protection of vulnerable persons from, for example, grooming,[3] program managers may need to ensure that inappropriate language is flagged quickly. But what about conversations between adults (for example, students at different years in the same university)? This is very much a grey area. On the one hand, the program should not be used to disseminate racist, homophobic, or other offensive comments. On the other, there is no reason why this medium should be singled out for specific attention. Common sense suggests ground rules for all participants along the lines of:

- While the program does not wish to suppress participants' views (and indeed, wishes to promote different perspectives), being part of the program brings with it a responsibility to avoid giving gratuitous offense.
- Should you consider a comment inappropriate, you have a responsibility to query the person's intent and to make them aware of your concerns. In most cases, the matter will be no more than a momentary thoughtlessness, or a misunderstanding.
- If the problem persists, with repeated use of inappropriate language, then you should consult your coaching/mentoring supervisor or programme manager.

A technology issue in process management is whether or not to archive e-mails. In young person mentoring, this is advisable to create an audit trail, if there are concerns about a particular relationship. In other programs, it may be simpler and more appropriate to allow participants to decide for themselves what to archive and what to delete.

Another often omitted process is how relationships will be ended. Good practice is to ensure that each pair conducts an open and honest review of what has been achieved and how, and that there is an opportunity for collective celebration, in programs where cohorts begin and end the formal relationship together. Holding an on-line party poses some difficulties, but some programs have made a point of rewarding participants through small gifts or publicity in their employee newspaper or e-zine.

TRAINING

Most programs require coaches/mentors to be trained in the role. The ISMPE standards also expect mentees to be trained, and good practice generally recommends that program coordinators and steering group members also receive appropriate training. The more people understand how the virtual learning process works, the more easily they can contribute to its success.

Our case studies include successful examples of both on-line and face-to-face training. A frequent criticism of on-line training is that coaches/mentors miss the opportunity to learn by practice—so they have knowledge but not necessarily the skills to put it into practice. It's true that face-to-face training allows for feedback from observers, who may be co-participants or trainers, and this can't be replicated easily on-line. However, if we break away from the assumption that training has to be delivered in one lump, a variety of on-line possibilities open up. For example, participants can practice on colleagues or program staff and then engage in co-coaching conversations about their experience, which can then be shared amongst and discussed by the learning set. This approach requires a skilled moderator, but it can be highly effective. It has some advantages over face-to-face training in that the learning elements can be broken down into small chunks and spread over several weeks or longer. This gradual build-up of skills is much more like "natural" learning and hence more likely to embed.

A variation recently developed by one of the authors provides line-manager coaches with an on-line learning resource, in 16 modules. They can choose to work through the modules at any time they wish, when they can assemble their direct reports either face-to-face or synchronously on-line. These real or virtual meetings are an opportunity to discuss the module content and to apply the learning to coaching in their own team context. Those modules that include skills practice can also be continued asynchronously.

Some of the published guidance we have accessed insists that coaches/ mentors are trained separately from coachees/mentees. In fact, there are strong arguments both for training separately and together. The primary issue is not "are there things we want to tell one group and not the other?"—that is a recipe for suspicion—but whether one or both groups will feel constrained by the presence of the other, and hence be less open. (In an on-line context, mentors may sometimes be reluctant to admit personal weaknesses or fears outside of their peer group.) If that is not a problem, then there is much to be gained from having coachees/mentees explain their views to coaches/mentors and vice versa. If some differentiation is required between coach/mentor and coachee/mentee training content, our experience has been that this can easily be managed by expecting everyone to read and discuss materials, but for only one group to carry out some specific exercises or skills practices.

A major mistake made in many programs, both face-to-face and virtual, is to assume that training requires no more than an initial sheep-dip. In reality, people take time to absorb ideas and develop skills, so international standards expect there to be a series of training events (which may take the form of individual or group supervision) over at least the first six to twelve months. A useful option, particularly suited to virtual programs, is to time input according to the key transitions that the learning relationship goes through. In mentoring, the transitions occur between the following phases:

- Rapport building (at the beginning)
- Direction setting (creating a sense of purpose, and gradually clarifying goals)
- Progress making (the core phase of the relationship)

- Winding up (bringing the formal relationship to a clear end)
- Moving on (achieving a more ad hoc, friendship-based relationship, or finding a new mentoring partnership)

Coaching phases follow a similar pattern, consisting of:

- A brief *contractual* phase, in which coach and coachee define the objectives of the relationship and whether they can work together
- A *transactional* phase, in which the coach helps the client explore specific issues of behavior, leadership, or skill
- A *review* phase, usually short, in which they assess the outcomes of the relationship together and, where appropriate, with input from other sources.

If the virtual coaching/mentoring program has an efficient monitoring system, it should in most cases be possible to provide participants with on-line learning materials and discussion opportunities more or less just-in-time to manage each of the key transitions. The opportunity to practice therefore becomes part of the relationship development.

An issue often not addressed in program planning is "what happens to participants after the program?" Do they move on to other learning partnerships? If so, can they continue to access the program's on-line resources, even though they may not be active participants in *this* program any more? Are there opportunities for mentees/coachees to become coaches/mentees, and for coaches/mentors to build on their experience to achieve further qualifications?

Not every coach or mentor wants to acquire a certificate or diploma in the role—no more than line managers want a financial qualification—but it can be a strong motivator for some people. There is now a wide range of qualifications that can be customized to the specific coaching/mentoring program. However, there is also a wide variation in quality, so it is advisable to seek a qualification that is accredited to a reputable international standard-setting body specializing in the coaching and mentoring field.

The content of training will vary according to the program purpose and stakeholders. For example, while a university-located mentoring program for postgraduate students might put a fair amount of emphasis on concept and theory, a program for disaffected youth that uses older peers would tend to focus more on practical techniques. However, all training should at a minimum cover:

- The definition of coaching/mentoring applied in this context an the purpose of this programme
- Roles and responsibilities

- How to manage the relationship (both when it's going well and when it encounters problems)
- Basic skills and competences required of participants
- How to manage the technology
- Codes of conduct and other guidelines
- Where to go to for help outside the relationship
- Other learning resources available

It goes without saying (but we'll say it anyway) that the trainers must be competent e-mentors/e-coaches in their own right.

MONITORING AND EVALUATION

In many ways, monitoring and evaluation are easier with virtual coaching/mentoring than face-to-face, because:

- There is (in programs with centralized message management) a ready record of the frequency and length of e-mail interactions between participants.
- With everyone on-line, questionnaires become part of the program process, so response rates tend to be higher.

Good practice includes:

- Establishing a measurement baseline at the beginning of the relationship. This might cover:
  - The expectations of each party with regard to each other and the learning outcomes
  - Current skill level in coaching/mentoring (or being coached/mentored)
  - Current skill levels in any specific area targeted by the learning relationship
  - Organizational or societal outcomes (e.g., improved employee retention amongst participants, or reduced recidivism amongst ex-offenders)
- Reviewing relationship quality after 4–6 months and then every subsequent 4–6 months (e.g., have expectations of behaviors been met?)
- Reviewing overall relationship progress. (How frequently have they had virtual meetings/conversations? What phase of relationship evolution have they attained?)
- Assessing outcomes, for participants (both parties) and for the organization/society.

One of the authors designed an on-line resource to carry out this evaluation process in the context of developmental mentoring, using questions double-validated in an extensive research project.[4] The resulting database allows program coordinators both to monitor what is happening in their program and to compare results against a basket of other organizations' programs.

While evaluation typically takes place at preset times, monitoring is usually continuous, so the speed of response should be relatively rapid.

Other useful questions relating to evaluation and monitoring are:

- Do we have a clearly defined process for quality improvement? (i.e., how we will make use of evaluation data?)
- In what ways do we need to demonstrate value-for-money?
- Can evaluation be undertaken using internal resources, or do we need to have an independent process?

HOW TO ESTABLISH A VIRTUAL DEVELOPMENTAL RELATIONSHIP OUTSIDE ORGANIZED PROGRAMS

Finding a coach or mentor on-line is not difficult. Finding the right one may be more problematic. The key steps are:

1. Decide what you want an e-coach or e-mentor for. What can they do for you that a book or evening class can't? What's the advantage of an e-pairing over face-to-face?
2. Decide what kind of person you want to work with. How will their experience differ from yours? Their perspective and knowledge? What kind of personality would give you the right balance of being able to build rapport yet provide sufficient stretch in the relationship?
3. Capture these thoughts and decide the timeframe over which you want the relationship to function.
4. Go hunting! Ask your friends and colleagues for recommendations. Place a "Coach/mentor wanted" ad in your blog. E-mail potentially suitable people you encounter in magazines or who have written articles you find on the web. Post requests on websites such as Plaxo. Join Second Life and let your avatar go shopping for an e-coach/e-mentor. In other words, use your ingenuity and the power of the web.
5. Interview the most likely-sounding of prospective respondents, either by phone or e-mail. Check out their own websites. Narrow the choice to two or three and interview them again. Let them demonstrate their coaching/mentoring style by giving them a small issue to guide you on. When you make your choice (and you can choose more than one, if you wish and have time for so much correspon-

dence), keep in mind the key question of balancing rapport and stretch—can you see yourself still being stimulated by dialogue with this person in six months' time?

6. Tell the chosen mentor why you have decided they are right for you. Then tell the ones you turn down about your choice and thank them. Promise to keep in touch with them anyway—they may be right for you next time around, or will be useful conduits to your next e-coach or e-mentor.

7. Make sure you prepare well for the first and subsequent sessions. Determine to be open and candid and to reflect on what has been said in the learning dialogue. After all, you want to sustain this relationship, if it is useful to you.

PILOTING THE PROGRAM

With face-to-face programs, it is usually relatively easy to carry out a pilot at low cost, and to fine-tune for a full-scale launch. With e-coaching and e-mentoring programs, this isn't always so simple. For a start, part of the rationale for the project may be economies of scale. If you have to develop on-line application, matching and training resources, the cost of 20 participants is not going to be much different from that for 200. Moreover, the costs of making changes to the software can be considerable. However, there are some practical ways to pilot in stages. For example:

- Stage 1: small scale group of 10 pairs, selected by "tap on the shoulder" from people already known and matched by hand. This group is facilitated in working through the pages of the intended on-line training resource, in a classroom situation, and their feedback gathered. They communicate using unmonitored e-mails, but keep a log of their coaching/mentoring discussions.
- Stage 2: a group of 20–30 pairs, attracted by a publicity campaign and using on-line applications. Matches are again "arranged" centrally, but they are expected to select from up to three choices offered. On-line training is launched.
- Stage 3: a second group of 20–30 pairs, with the whole process IT-based, including programme administration and e-mail monitoring.
- Stage 4: full-scale launch

This particular example would not be suitable for young people at risk, where e-mail monitoring would need to be included right from stage 1, but the general principle of staggering the technology introductions does seem to be sound. If nothing else, it allows the program coordinator and his or her team to troubleshoot each element individually, rather than deal with problems across the whole process. Moreover, one of the golden rules of IT design is to make sure things work in the non-virtual world before trying to automate them.

WHAT GOES WRONG WITH VIRTUAL COACHING AND VIRTUAL MENTORING?

Some of the problems we have already seen—for example, insufficient preparation and funding shortfalls. Among the most common others are:

- Failure to secure sponsorship
- Unrealistic expectations of internal IT resources. In one multinational company with a very large IT department, the e-coaching and e-mentoring programs were announced, then delayed by nearly a year, because other software writing projects took priority. As a result, the program had an uphill struggle to regain credibility.
- Sheep-dip training—too little to give participants confidence to behave differently
- Bureaucracy. E-coaching and e-mentoring relationships require support and access to guidance when they need it. But they also want to get on with their conversations in their own way. In one program, participants had clamored for on-line resources about coaching so that they could keep on learning. At a review session, the training and development manager berated them that they had used this resource, on average, only three times each over six months. "Ah", said one of the coaches, "but the point is, it was there when I needed it."
- The "mentoring menopause." Many relationships experience a sense of deflation after six to nine months, when their discussions run out of steam. With help from the coordinator or a coaching/mentoring supervisor, they can usually regroup, focus the conversations on deeper issues, and regain their momentum. Many programs miss the signs of this phenomenon, or do not recognize it at all—hence their relatively high levels of drop out.

SUMMARY

In this chapter, we have attempted to review some of the key dynamics of the virtual coaching/mentoring program. Almost all of these dynamics relate to people processes and to general principles of effective management.

However, the other critical half of the story—the platform on which the people processes depend—is the technology. In the next chapter, we review how to get the best out of the IT solutions available.

NOTES

1. Clutterbuck, D and Ragins, B.R., *Mentoring and Diversity—An International Perspective*. Butterworth Heinemann, 2002.
2. http:// www.staffs.ac.uk/schools/sciences/ementoring
3. When a person tries to set up and prepare another person to be the victim of sexual abuse.
4. *Mentoring Dynamics Survey*. http://www.clutterbuckassociates.co.uk/content/Company/Products/MDSOnline.aspx

REFERENCES

Clutterbuck, D. (2006). How to be a great programme coordinator. Occasional papers. Clutterbuck Associates. Burnham, UK.

Clutterbuck, D. (2007.) *A longitudinal study of the effectiveness of developmental mentoring*. Unpublished doctoral thesis submitted to Kings College, University of London.

CHAPTER 3

THE TECHNOLOGY IN PRACTICE

Zulfi Hussain

In the previous chapters we have seen how virtual media have enriched and in some cases offer a complete viable alternative to traditional face-to-face developmental relationships. In this chapter, we explore the new technologies in more detail, examining how they can support the developmental relationship and what program designers and participants need to take into account in using them in a coaching and mentoring context.

THE CHOICES OF BASIC TECHNOLOGY

There are many different ways that technology can be used to facilitate virtual coaching/mentoring. Therefore, one of the first things to look at and check is the infrastructure you have in place. Then ask yourself, will the scheme be web or email based? Will it be a stand-alone scheme or are you adding this to an existing traditional face-to-face scheme? What other facilities does it need and/or is it likely to use?

Security, confidentiality and data protection are essential elements of any virtual coaching and mentoring program, so it is imperative that all

Virtual Coach, Virtual Mentor, pages 53–76
Copyright © 2010 by Information Age Publishing
All rights of reproduction in any form reserved.

reasonable measures are put in place to minimize risks of any potential breaches. Consideration should be given physical and logical security as well as safe backup, storage, and retrieval of data.

Using technology intelligently can add immense value and provide many benefits while considerably reducing ongoing costs for both participants and the program.

The Web-Based Model

Web-based models usually require the design and development of a new website. Most web designers will not have much understanding, if any, of the practicalities of coaching and mentoring, so you will need a detailed description of how you want to facilitate conversations between participants. At the minimum, you will need to define:

- Whether you want synchronous or asynchronous communication, or the option for both
- The expected number of users at any one time
- The level of security you require
- The nature of all interactions through the website between the program manager and participants
- What signposts you will want on the home page to other pages of content
- Any hyperlinks to other resources
- How users will register and once registered, what access they will have to their data
- What information you want to collect about use of the site and the various resources within it
- How much information you will hold, who will have access to it, and how
- How the website will be updated and by whom
- Your expectations in terms of how much maintenance the site will need.
- Whether or not you want the website to have audio and video capability rather than just text.
- A business continuity/contingency plan in case of the website not being available for any reason.

An alternative option is to use an existing web-based scheme through a third party, which can help reduce the cost, as this is a shared resource used by many different organizations and the development costs have therefore been largely written off. If you want special features, however, this may be expensive and even a relatively straightforward application may need some modification. For example, a system using American English may need language adaptation to make it suitable for users in the UK. The other option is to buy a customizable coaching/mentoring software package and integrate this into your own infrastructure. This latter option can potentially be more flexible but may cost more in the initial set up stages.

Once the site is up and running, the participants can access the web pages via the internet and/or intranet, use discussion forums, and even read or download other relevant coaching and mentoring resources.

Email-Based Model

Email is well established, but there are still some objections to its use on the grounds that it requires computer literacy. This is not a coaching or mentoring problem, but a skills problem that can be easily addressed with a little bit of training and practice. Contrary to common opinion, age is not a barrier to acquiring familiarity with email.

In formal email-based models, especially in schools and other similar organizations, the participants use a unique email address to communicate with each other. This helps with the additional security and monitoring of the scheme and its participants. This usually needs to have technology in place that provides a safe and secure environment for email exchanges and archives all email messages and tracks email communications between the mentoring pairs.

In other email-based models, participants use their normal day-to-day email address or set one up specifically for this purpose. A separate email helps to manage and track the coaching and mentoring emails more easily. With a dedicated email address (i.e., one that is only used for mentoring emails), you don't have to trawl through lots of other types of emails when you are trying to find and track correspondence.

The mechanics of using email are pretty straightforward. Typically, the coachee /mentee will set down in a message the issues that are important to him or her at that time, using software such as email that is designed to enable the receiver to insert text into the original message. The coach/ mentor can then 'thread' the responses between the coachee's /mentee's sentences. The threading is a really important aspect, since it allows the mentor or coach to highlight the elements of the text that are important, and insert the response next to the original text. This makes it almost the next best thing to having a conversation.

Ground rules, of course, still have to be agreed, just as they would in a conventional coaching/mentoring relationship (see the discussion on *netiquette* in Chapter 2). Instead of regular meetings, regular times for sending and receiving e-mail messages can be set up, or a more spontaneous approach can be agreed upon that provides flexibility for people who have busy schedules and differing work and lifestyles.

The Telephone-Based Model

The telephone technology has been around for a very long time, and people are comfortable with using it in any type of transaction (although many people have a "telephone persona" different from their normal conversational style). It helps increase people's availability, as they can operate at any distance and across time zones. It can also be used in a more impromptu way, or calls can been scheduled in at a fixed time.

Many individuals and organizations use telephone-based coaching and mentoring through the traditional telephone network, but more and more use is started to be made of internet based calling using Skype and other IP-based calling systems—not least because they are much cheaper than standard landline, especially for international calls, and free if using the same network (e.g., Skype to Skype calls).

As we discussed in Chapter 2, there has to be much more proactive listening in a telephone-based session, since all the visible clues found in the traditional face-to-face session are missing. Although some of the non-verbal clues are lost, the tonal variations, speed of speech, and even the silences can still impart vital clues to the participants.

The Videoconferencing Model

Although videoconferencing technology has been around for a long time, there has been a very limited use made of it for coaching and mentoring. One of the reasons for the limited use has been availability and cost. However, this is all set to change as Broadband is now very widely available and the use of webcams is becoming commonplace. Although originally intended for meetings involving more than two people, videoconferencing for a coaching or mentoring dyad provides a level of intimacy very close to that in face-to-face conversations.

There are now two main options for videoconferencing: desk-top and conference room. Packages like Skype and MSN provide low-cost desk-top conversations, with a picture of the learning partner on-screen. Low resolution and technology gremlins can be a problem, but for most people, the system is good enough.

With continuous improvements in Broadband speeds, falling hardware prices, and enhancements in hardware and user applications, this is fast becoming a very flexible, affordable, and attractive option.

Videoconference facilities using state-of-the-art technology provide a much higher level of definition, clarity, and intimacy, such that it seems the other person is just across the table. Using a separate conference facility also provides privacy, and users can exchange and work on documents together as they talk. There are many video conferencing systems available (e.g., Cisco Telepresence, Adobe Connect Pro, etc.), but they all have one thing in common: to provide an 'in person' experience so you can meet people without the need for travel, and communicate and collaborate more effectively anytime and anywhere. This saves time and money while also helping you reduce your carbon footprint.

Deaf and hard of hearing individuals have a particular interest in the development of affordable high-quality videoconferencing as a means of communicating with each other in sign language. This opens up a whole new community of interest for virtual coaching/mentoring.

Rapid technological advances will improve both of these approaches. Desktop conferencing will permit full-screen pictures of the other user(s), and we will soon even be able to achieve 3D effects in the teleconference suite.

ENHANCING THE CONVERSATION WITH OTHER TECHNOLOGIES

One of the benefits of the e-conversation is that it can be more than one conversation. While participants are exchanging verbal or written thoughts, they can also be sharing pictures, weblinks and documents that expand on the core conversation. For example, a recent but powerful addition in virtual coaching and mentoring is the use of the Wiki to share resources and materials. The Wiki allows users to freely and relatively easily create and edit web page content using any web browser (such as Internet Explorer, Netscape, etc.). The beauty of a Wiki is that it is simple in concept and allows non-technical users to create and use the content. It is a great enabler of collaboration and sharing.

Much simpler, and much longer established, is the Learning Log. Essentially a personal diary, it is used in many coaching and mentoring programs to assist participants in reflecting on their learning and recognizing how far they have come on their learning journey. In some academic environments, the learning log may also be shared with tutors and form part of the course assessment.

Technology is a great enabler that helps overcome time and distance barriers as well as helping the process of understanding cultural differences and norms. Communicating across cultures can be challenging at the best of times, but especially at a distance. Therefore, using videoconferencing facilities enables the correct reading of verbal and non-verbal cues and signs. It also provides context and sub-context, which are essential for more effective understanding.

USING THE TECHNOLOGIES

This section lists of some of the most frequently used virtual technologies that can be used for e-mentoring and e-coaching, with more detail on where they are most effective.

Landline Telephone

A traditional landline telephone system commonly handles both signalling and audio information on the same twisted pair of insulated wires: the telephone line. Although originally designed for voice communication, the system has been adapted for data communication such as telex, fax, and internet communication. This form of technology enables direct communication between e-mentors and their learners in real time and is easy to connect to—you just need to pick up the phone and dial the corresponding number. Confidentiality is easily assured as you can only connect to the number dialed (unless there is more than one phone in the household where someone can pick up another handset).

Advantages/Disadvantages
This form of communication is easy to use and easy to get hold of. It is still common for most households to have a landline telephone, and you do not have to worry about connection problems. It can be argued that it is easier to use for communication compared with email or text services because questions and answers can be exchanged in real time as opposed to having to wait for a reply. It can also work out cheaper than using a mobile telephone as well as possibly cheaper for transport costs incurred for face-to-face interaction.

However, because it is a fixed line, in can be an inconvenience to arrange when to both be at the phone at the same time and, particularly with younger individuals, speaking on the phone can feel quite intimidating compared with emails. It is also seen as less exciting than the current technology available!

Mobile Telephone

The mobile phone (also known as a wireless phone or cell phone) is a short-range, portable electronic device used for mobile voice or data communication. Nowadays, mobile phones may support many additional services including SMS for text messaging, email, access to the Internet, Bluetooth, infrared, and camera with video recorder and MMS for sending and receiving photos and video.

Advantages/Disadvantages
As it is portable, contact can be made easily at anytime—as long as there is signal and battery. A pay as you go handset can also be bought for as little as £9.99. The text messaging service can also be used to communicate messages, although this is time consuming and restrictive to small messages. Texting is also expensive compared with email, unless you use infra-red or Bluetooth—but this can only be used when in close proximity to the person's phone. Video messaging is also a good way of communicating, but this can be done more easily and cheaply using different methods.

Using a mobile is less reliable. The battery needs to be sufficiently charged, and there needs to be a signal for it to work. The quality is sometimes not as good as a landline. It can also prove costly if the mobile plan is not the correct one for the specific needs of the owner. However, mobiles are generally more convenient than landlines for reaching people, and nearly everyone has one. In some developing countries, they are the only practical way of reaching participants. They are also easy to use and come with detailed instruction books. Staff at the shops where they are bought can also give demonstrations in most cases.

Email

Email is a common communicating device today. It uses an Internet connection to send mail instantly. An email account is easy to set up—it takes about five to ten minutes, and there is a wide range of providers (Hotmail, Gmail, Yahoo, AOL, GMX, Inbox.com, and Zenbe, to name a few).

Advantages/Disadvantages
Email can be used to send messages of any length to mentors and mentees. Communication can therefore be maintained without having to travel

to meet face to face and as message transfer is instant, it is a fast and easy way to send information. It is very easy to use and set up, and most email providers include an extensive help section and troubleshooting option.

Email is also good to use from a financial perspective. It is free to send email, whereas in most cases you would have to pay for a phone call or travel expenses. You do, however, need an Internet connection, which may cost, but there are some free Wi-Fi hot spots in some libraries, cafes, and hotels that can be used.

Email has the benefit of being able to send as much text information as you like, including video and picture messaging, although the time in which you can send and receive these types of messages may increase and some servers may not be able to handle sending large files. Communicating in this way may seem less intimidating than talking face to face, which could benefit some mentoring partnerships, and email accounts can be set up to be accessed from any computer, making it very accessible.

One of the main problems with emailing is keeping up to date and answering messages. Emailing may prolong the mentoring process by having to wait for replies, whereas answers can be achieved instantly on the phone. It is also less intimate as you cannot see the person you are talking to, which may not appeal to some coaches/mentees. In order to use emailing effectively, accounts must be checked regularly and emails replied to as soon as possible so that problems or questions that have arisen in an email get sorted almost as effectively as they would on the phone.

Confidentiality is easily achieved with email. If the address is correct, then it is not normally possible for others to read it (unless there is a virus or someone has hacked into the account, which is rare). The exception to this is where emails are monitored, say within an organization. In this case emails could be viewed by authorized individuals within the IT department. There are usually stringent guidelines within organizations for monitoring emails and/or calls in line with data protection guidelines.

MSN Messenger

MSN Messenger is an instant messaging service. It is similar to email in the way it works, except the recipient receives and replies to the message as if it were in discussion.

Originally it started off with the exchange of text, but more recent versions of MSN Messenger (windows live messenger) can include video conferencing, the uploading of webcam images, exchange folders for picture and video, and preloaded animations for animating the text you write.

It is simple to use although a hotmail email account is needed to log in with. This can be created in a matter of minutes at www.hotmail.com. Messenger can be accessed in two different ways—either go online and download the latest edition of the MSN Messenger software, or achieve instant access through the web version of the software (at http://webmessenger. msn.com/). The web version is only basic, but means that the messenger can be accessed at any computer.

The downloaded version (which can be found at http://get.live.com/ messenger/overview) can include all the features like video conferencing and webcam use, although, once it is downloaded, it can only be accessed on that computer.

The login details (which are the email address and password) need to be input before being able to access the account. When login details are correct, a new window opens and it shows a list of 'friends' who are online and offline. 'Friends' are other MSN users who have been accepted as being allowed to communicate with that person on their account, and unless they enter into a free 'chat room' where anyone can enter, no one other than verified 'friends' can communicate. Friends are either verified by sending out a request to be a friend (by typing in an email address of a friend so they can confirm) or by receiving a friend request (where a small box appears and the option is given to accept or decline a friendship requested by the other party).

Communication can begin when an online friend is selected. A new window opens up and the option to 'send message' appears, which is the main basis for the communication.

Advantages/Disadvantages

MSN Messenger has the main advantage over email in that it is as instant and non-threatening as email but questions and answers can be exchanged without having to wait for the other person to check the email—therefore saving on time. It is also free to use except for needing an internet connection, which may cost. MSN is also very popular with teenagers and young adults and saves on travel costs. It does, however, take slightly longer to have a conversation because of the need to type as apposed to just speak but it may be more exciting to use than just the telephone. The latest software also includes other ways of communicating that overcome these barriers.

If one or both of the participants in a conversation have a webcam, then this can be used to enhance the experience as though talking face-to-face. This requires a good internet connection and enables the person receiving the webcam to be able to see the person to whom they are talking. This can also be taken one step further by starting a 'video call'. This requires a microphone of some sorts for inputting sound data. When a video call is in progress, one or both parties (depending on whether they want to broadcast one, want to accept a broadcast, or whether they have the right equipment) will be able to look at and speak as though in a normal conversation. The option to transmit just voice is also available and files can also be sent over such as Word documents, pictures, and presentations.

The video call would sound like the perfect answer to e-mentoring as the face-to-face contact is there, except there is no need to travel and it is easy to set up. However, the quality is not perfect and there can be long delays when speaking and receiving. Confidentiality is easily maintained in any method, however, as only selected friends can view the conversation. Conversation can also be set up to have more than one person in (except for videoconferencing), but again, they can only join in when invited.

MSN also includes a range of troubleshooting and help pages as well as online tutorials to help aid the learning process and to sort out any problems. MSN is very easy to use and to get used to, and if for any reason the MSN Messenger doesn't work, the hotmail email address is always there to use instead.

Skype

Skype is another way of communicating online. It is downloadable software which allows a person to buy credit and speak to other Skype members for free, but to also communicate with landline phones and mobiles which costs considerably less than from a landline or mobile. It also offers the option of having video calls and instant messaging as well as SMS messaging to mobile phones cheaply. Voicemail is also available and when a contact is not online, their calls can be forwarded to their mobile or landline telephone.

Downloading Skype is free (at http://www.skype.com/intl/en-gb/download/skype/windows/), but a microphone and speakers are needed for video conferencing. Other requirements include a PC running Windows 2000, XP or Vista (Windows 2000 users require DirectX 9.0 for video calls); internet connection (broadband is best, GPRS is not supported for voice calls, and results may vary on a satellite connection); and a computer with at least a 1GHz processor, 256 MB RAM and a webcam.

Skype is similar to MSN in that communication can only be made with contacts that have been verified by the user as being 'friends,' which makes conversations confidential. What is better is the ability to use this software as a telephone system, and users can even purchase wireless phones to make it feel more like a telephone call. The user can also make or receive a call involving more than one person if needed and, if for some reason the contact they need to speak to is not online, they can easily make an ordinary telephone call or text message to the intended recipient's telephone.

Call quality is good. Skype has taken extra measures to make sure that the sound is as much like a normal telephone as possible. The quality of a video call depends on the quality of the webcam and webcam software. Skype is easy to use and for telephone type calls, the connection to the call is the same as how a normal phone is used. A number can be dialed to any country and the sender or recipient hangs up whenever they want. Instant messages can be sent in the same way as MSN; text SMS messages are sent by typing the message and sending to a phone number. Video calling is requested by one party and the option to join in is given to the other, but only two people can be in a video call conversation.

Advantages/Disadvantages

Skype can be used to benefit virtual coaching/mentoring in a variety of ways. The first being that it is cheap to use and to set up. As software, it does not contain one way of communicating, but many, so an e-mentor or e-mentee has the ability to select from a wide range of communication methods. The instant messaging on Skype can also act as an email type service, as a message can be sent that will appear when they next log in. Confidentiality is maintained by the ability to choose whom to talk to, and participants can always clearly see to whom they are talking. The videoconferencing is better quality than the MSN videoconferencing, and the ability to ring phones is like using an ordinary phone. Time and money can be saved on travel, and with so many ways to communicate, it is easy to keep in touch.

As an e-mentor or e-mentee, all types of communication included in Skype can be used to their full potential. The instant messaging service for sending a quick short message, a phone call to have some questions clarified, a video call for a regular general catch up and SMS text services reminding the e-mentor and/or e-mentee that they are meeting online in 10 minutes.

Like with most software, there is also extensive online help and troubleshooting. There are also alternative sites like Twitter who offer the same services. Skype, however, remains the most popular at this point.

Moodle

Moodle is a free software e-learning platform. It is designed to help educators create online courses with opportunities for rich interaction. Its open source license and modular design mean that people can develop their own pages and plug ins such as activities, resource types, question types, data field types (for the database activity), graphical themes, authentication methods, enrolment methods, and content filters.

In order to participate, the user must create a new login account, and then download the software—this is all free. The participant can then begin exploring what is offered. As well as having the ability to design online courses, the mentor can also create informative wiki pages (internet pages where the content can be changed by anyone as long as it is approved). It also proves to be a great resource for teachers in general with lots of pages about effective learning and information about how to make the most out of Moodle.

It does, however, take some getting used to, especially when beginning to design pages and learning how to use the code, but there is an extensive range of help resources including online discussion, training videos, online documentation, purchasable books, and help features. If the mentor is willing to sit down and get used to it, it may prove to be an invaluable resource for teaching.

From the mentee's point of view, this site could be a useful reference point to look at documents that the mentor has produced, as well as other pages of interest found in Moodle. The mentor can even design pages specifically for their needs in the format of an online quiz, for example, which would increase participation and create a better learning scenario than simply reading an informative wiki page. The mentor can also create a whole course that the students can take in their own time.

Advantages/Disadvantages

Moodle has several benefits over face-to-face learning. It can be accessed at any time and the help resources and online databases make solving problems relatively straightforward. It is also aesthetically attractive, which makes interaction more interesting and fun. The interactive side of things enables the learner to be engaged in a way that traditional teaching does not.

In conclusion, the software sounds like a brilliant way to aid learning, but it is complicated enough that it may take a while before the mentor has the ability to start designing their own courses. From a mentee's perspective, there are thousands of articles, lessons, and courses already on the site that may prove relevant to their learning. The help features are also very good which means that all the resources needed to start designing are readily available.

Adobe Connect Professional

Adobe has created a conference call software with the ability to create real-time meetings, seminars, screen shares, image share, and file exchange, in addition to hardware such as interactive whiteboards for both business and educational purposes.

A range of packages are offered, including Adobe Acrobat Connect professional (which includes the installation of interactive whiteboards); Adobe Presenter (for use with PowerPoint to create narrated, self-paced e-learning courses and presentations; with quizzes and videos); Adobe Connect Training (for managing e-learning courses and curriculum, including reporting capabilities at both the participant and course level); and an Adobe Connect Events module (for large, online presentations up to 2,500 participants, with user registration, automated reminders, and tracking capabilities).

In effect, all these technologies enable learning from a remote location, may it be a quick meeting or a lesson. Adobe Connect Pro allows a mentor to create and deliver self-paced courses, conduct highly interactive virtual classes, and efficiently manage training programs. The software can create high-impact content, deliver courses via virtual classrooms or self-paced courses, reuse learning assets with templates and content libraries, and track course effectiveness with robust reports.

This software, like Moodle, can be a great asset to e-learning but does appear more user-friendly. It is clearer on how lessons can be created and made interesting like how to input video or audio, and it can easily be done through Microsoft Powerpoint and then uploaded. Students also do not need software apart from Adobe Flash in order to participate.

Virtual classrooms can also be created, which can include live audio and video streaming, interaction like in a physical classroom, and quizzes where the results are recorded. Break-out rooms for discussion can also be created, among many other interesting learning devices. There are already courses available from the content library to teach within the virtual classroom, and teachers can teach the entire course or allow learners to go through it at their own pace. Models are also included where lessons can be specifically designed in addition to what already exists.

Advantages/Disadvantages

There are many benefits including making it more convenient for working from home, avoiding traveling, and being able to keep a good record of things that have been taught/learned. Additionally, it provides a good structure for learning if used correctly and has the potential for injecting interest and excitement into a lesson. It may be difficult to get used to the technology initially, but once it has been used a few times its benefits will become apparent. The fact that Moodle is free makes it very attractive, but the Adobe technology is more easily managed. There is a lot of online help and troubleshooting as well as a 24-hour call center available to answer needs. However, it should be mentioned that all lessons taught with technology should have a back-up plan; there are too many things that could go wrong on any given day that could halt the lesson. This is why it is handy that some of the resources are available all the time as apposed to just having live lessons. Students can work on these lessons in their own time and avoid meeting times and dates where the technology must work for live sessions.

Other Technologies

There are many sites on the internet that focus on social networking and communication. Many of these would be relevant to e-mentoring in that they are all tools that can help contribute to better communications between the mentor and the learner, but in most cases the only communication available is through emailing, posting pictures and video, and posting text. Many social networking sites—the most popular being Facebook—include this basic model. Other popular sites include Badoo, Bahu, Bebo, Perfspot, and Myspace.

Other social networking sites to note include places like Habbo and Second Life. These websites involve the participant creating a virtual profile of themselves—it could be anything or anyone. Particularly in Second Life, people can actually live out their 'second life'. E-mentoring can play an important part in Second Life, and people can even earn a living by teaching in the virtual world! This is done through posting lessons, online quizzes, and information about different topics as well as online learning forums.

LiveJournal is an online database of journals. On this site, a mentor can publish documents relating to the teaching topic (or other), and the student can subscribe to the feed. Whenever a new document is uploaded, the student can be notified and read the document. It can also be used for emailing and chatting in a confidential space.

Sites like Stickham and HelloWorld are based around broadcasting films. These can be anything from dancing around your room to a lesson. This may be helpful in e-mentoring, as the student will be able to watch the lesson online in real time, as long as they know the feed address and the time that it will be broadcast. The problem with this is that it is not interactive and anyone can see the tutor giving the lesson if they wanted to. On the other hand, it can be very useful for skills-based coaching and group coaching.

DESIGNING THE TECHNOLOGICAL SOLUTION

Given such a wide choice of media, the program manager will need to give considerable thought to the best delivery method for the particular coaching/mentoring application. The optimum solution will depend on a variety of factors, of which the most common are:

- Cost—what will our budget allow?
- Security—how important is it to monitor conversations?
- Quality—what level of service and system availability is required?
- Business Continuity—what sort of contingency do we need in case of system failure?
- Accessibility—how do we grant and monitor access to prevent any unauthorized access attempts?
- Gathering Information—what do we need to monitor and capture to help us evaluate the success of the program?

Table 3.1 is a helpful way of looking at the choices. Using informal technology means that the program relies entirely on publicly available resourc-

TABLE 3.1

Practical issues	Informal (open) technology	Supported technology	Managed (custom) technology
Accessibility	• Relatively easy access • Needs registration • Available to everyone.	• Easy access • Needs registration • Available to selected people	• Easy access • Needs registration • Available to selected people
Analysis, monitoring and gathering feedback	• Limited in terms of gathering and analyzing data.	Gathering and analyzing data tools are often built into the system, making it easier for analysis.	Customized data gathering and management tools are built into the system from the outset for ease of use and more in-depth analysis.
Confidentiality and security	• Reasonable levels of security and confidentiality.	Good levels of security and confidentiality are built into the system.	Good levels of security and confidentiality are built into the system.
Linking into other resources (e.g., information resources, learning logs)	• Fairly limited links to other resources.	Good links to other resources are provided by the system.	Customized links to other resources are readily available.
Closed sub-groups (e.g., mentor forum)	• Some systems offer closed user group facilities.	Closed user group facilities are available more often than not.	Closed user group facilities are built into the system from the outset.
Cost	• Relatively low and often free of charge to use.	There are initial development and setup costs that vary depending on how much work is required to be done from the outset.	Initial development and setup costs can be high but the systems are flexible and can be customized to meet specific needs.
Technical support	• Limited amount of help and support is available in terms of registration and technical problems with the system.	Good level of support is available as part of the overall service level agreements	Good level of support is available as part of the overall service level agreements.

es, such as email and telephone. Matching will largely be done by hand and the website, if there is one, will be simply a shop window for the program. Supported technology provides users with some discrete tools and information resources, tacked onto an existing intranet or other web-based resource. Managed technology involves an integrated system, from participant acquisition to monitoring and measurement.

The Rise of Mixed Media Coaching and Mentoring

The earliest applications of virtual coaching and mentoring were overwhelmingly single technology. Most current applications tend to involve multiple media—partly because there are more media available, and partly because program organizers have recognized that participants have different preferences and circumstances in how they find it easiest to communicate.

Critical issues in planning the technology mix include:

- What technologies do participants have access to?
- How comfortable do they feel in working with each technology?
- How homogenous are their circumstances? (The more varied, the wider the range of media that may be needed.)
- What is the cost of each and who bears the cost? (If participants have to pay for expensive media, they are less likely to use them.)
- What sort of training and support do participants need?
- What are the costs versus benefits in the short, medium, and long term?
- How do we future-proof the investment?
- What is the payback period and/or rate of return for the investment?

The Importance of Training

One of the recurrent lessons from our case studies is that participants need to have at least a basic competence in the technologies involved. Age, education, and geographical location will all have an influence on people's speed of assimilating a new technology. So will social factors—for example, texting appears to be more effective with young participants, because they already use it regularly. If a technology is only used by participants in the context of their coaching or mentoring relationship, they are likely to be less adept than if they use it more widely.

Effective training covers both process and content—how to use the technology and how to adapt communication to the technology. Key questions include:

- What level of ability do participants already have in using the technology? Is there a wide variation in level of ability and experience using this technology?
- What is the most effective way to deliver training? People with low computer literacy may not best be reached through computer-based training.
- What changes in communication skill and behaviors will be needed for participants to make effective use of this technology (mix)?
- What one-to-one support are we able to provide to participants? (For example, on-line or telephone help line.)

The Technology Strategy

Programs using supported or custom technology involve expenditure in software and hardware. As with any IT project, a technology strategy is essential to ensure that costs are contained and that the technology delivers what it is intended to in a sufficiently user-friendly manner.

Key questions to consider include:

- What computer equipment do we need? (Can we make do with existing equipment or do we have to buy new?) The prices of hardware continue to fall while the quality and functionality continue to improve. Therefore, the dilemma is when to invest in new equipment and whether or not this will give any added benefits or efficiency savings.
- What software do we need? Will off-the-shelf solutions give us what we need? Or do we need to create our own? Software design and developers continue to create new and improved applications all the time. Unfortunately, many of these do not meet the needs of all their users. This is due to the fact that users have varying needs and 'one size' does not fit all. Hence a balance has to be struck in terms of cost and benefits of a custom solution versus one off-the-shelf.
- What is the risk of unexpected downtime? What impact would downtime have? How can we minimize that impact? A full risk assessment should be carried out to identify all the risks and the associated contingencies in the event the risk is realized.
- What policies and procedures do we have to provide back-up? It is crucial to have proper back-up and restore policies and procedures in place that are tested regularly to ensure that they will work in case of system failure.
- What level of security do we need to have? (Not all programs need a high level of security, but the technology costs rise steeply with the

level of security required.) Good security provisions are absolutely essential to minimize any risk of breaches. However, a balance has to be struck between the level of security actually required to what may be just a wish list. This will help ensure adequate protection at a reasonable price.

Managing the Security Issues

Managing the security issues is absolutely essential. This is done by a process of preventing and detecting unauthorized use of your computer. Prevention measures help you to stop unauthorized users (also known as "intruders") from accessing any part of your computer system. Detection helps you to determine whether or not someone attempted to break into your system, if they were successful, and what they may have done.

There are three fundamental pillars that are seen as the foundation stones of all secure private communications known as the CIA of security. CIA stands for Confidentiality, Integrity, and either Authentication or, in some cases, Availability.

> *Confidentiality*—ensures that the conversation remains private (i.e., confidential). In other words, information should only be available to those who rightfully have access to it.
> *Integrity*—ensures that the contents of the message remains free from interference or corruption (i.e., the message remains intact). Therefore, information should only be modified by those who are authorized to do so.
> *Authenticity*—ensures that all the parties to the conversation are verified and are who they say they are.
> *Availability*—information should be accessible to those who need it when they need it.

There are many products available on the market that help protect your computer and information and give you a secure mentoring experience. The three most common internet security products are as follows:

> *Anti-Spyware Program Software*—Adware, Malware, and Spyware are not only potentially harmful to your computer's operating system, but they can be used to steal personal information. Anti-spyware programs that are designed to protect you against these threats should be used for protection.
> *Anti-Virus Programs*—These programs are designed to protect your computer against Spyware, Trojan horses, viruses, and worms that

can install themselves, through deceptive operations, onto your computer, and then do serious damage in the process. Anti-virus programs are designed to protect you against these programs so that you don't run the risk of your computer being ruined from the infection they create.

Firewalls—These help to ensure security while on the internet by blocking the entry of unauthorized traffic through authorized ports on your computer. Typically, a firewall will block or restrict access to the following:

- e-mails
- programs
- webpages

Any of these can create potential virus risks that threaten your computer's well-being, so a firewall is good protection against these issues.

One of the fundamental and very effective ways to protect your information is to create and use a password. When you are creating your passwords be creative and think outside the box. Don't use your name, birth date, or address in your password. Additionally, never use obvious things like the word 'password' or the sequence 1, 2, 3, 4. Your password should have seven or eight characters, because the longer it is the more secure it will be. You should choose something that would appear to be random to someone else and also use upper and lower case letters, along with characters to make it more secure and harder to crack.

- What level of security do you need in different circumstances?
- How do you prevent spamming, unauthorised access?
- What systems can you include to prevent inappropriate content?
- What questions should the programme manager be asking?

What Goes Wrong in E-Coaching and E-Mentoring?

Many of the problems we have observed with e-coaching and e-mentoring are common to face-to-face relationships as well: failure to ensure participants fully understand their roles, not managing expectations, over-bureaucratization, poor management of the program, lack of support for participants as they progress through the relationship, and inadequate monitoring and continuous improvements to the program.

However, there are also problems that apply only to virtual coaching and mentoring. These include poor system design that makes the system difficult to use and/or limits its functionality; weak security provision that could lead to potential confidentiality breaches; inadequate training that prevents users from being able to fully utilize the facilities; limited user access that leads to frustration; and poor netiquette that leads to poor communication and misunderstandings.

The Technology for Matching and Supporting E-Coaching and E-Mentoring

Informed observers of coaching and mentoring are sometimes highly critical of on-line matching processes. Garvey (2004) points out that "quick fixes" are unproductive. Matching solely on demographic factors or on the basis of a specific experience gap can be overly simplistic and does not accurately reflect the complexity of the learning relationship.

A couple of decades ago, one of us carried out some ad hoc research to understand how HR professionals determined who would be a good match for whom. The conclusion of these interviews was that they looked first for compatibility (will they get on together?), then for learning potential, which often meant significant points of difference. Mentees with relatively high socio-emotional and learning maturity can typically accommodate a higher degree of difference (the grit in the oyster!).

Almost all matching software currently available is a one-step process, simply matching on mainly demographic criteria. A more robust approach is taken by matching in two steps. The first compares mentor and mentee on the basis of compatibility of values, using the 16 factors defined in Reiss values profiling (Reiss, 2008). Values compatibility has been shown to be the most significant predictor of rapport in the dyadic learning relationship (Hale, 2000a, 2000b). This initial sorting ensures that there is sufficient appropriate similarity between mentor and mentee for them to build trust and mutual respect at an early stage of the relationship. The second step sorts by demographic factors (age, location, gender, academic discipline, etc.) and other areas, where difference of experience will create significant opportunities for learning. As in most other on-line matching systems, the demographic factors can be weighted in each search.

Our experience of effective matching in other contexts also leads us to conclude that offering several choices (three appears to be the optimum) allows the learner to feel in control of the process and gives the mentor greater self-confidence in the knowledge that they have been selected by the mentee, rather than just by an electronic system.

Other practical lessons from introducing a matching system include:

- Pilot it rigorously on a control group first—do the matches that emerge really meet the criteria you set?

- Test the automated results against your intuition, as to who will work best with whom. If there is a significant gap, investigate.
- Involve participants in the evaluation of the matching pilot and in the design of the matching criteria.
- Be very clear in your own mind and articulate within the program what are the core elements of similarity you are looking for and the core "stretch" characteristics, where difference will be beneficial.
- Build in a system for analyzing those matches that do not work well. Are there recurrent combinations (similarities or differences) that are more commonly associated with poor quality relationships?

Support technology for e-coaching and e-mentoring is often weak, although some customized sites, such as Triple Creek, do provide on-line advice for both participants and program managers. Support systems, which you may wish to consider integrating into your on-line provision, include:

- *Budgeting software* to allow the program manager to control expenditure (particularly important if the number of applicants exceeds expectations, or if there is an unexpectedly high turnover of coaches/mentors)
- *An information database.* One of the authors has recently developed the first on-line encyclopaedia of coaching and mentoring, called CAMeO,[1] as an off-the-shelf repository of information on coaching that can be searched by keyword, alphabetically, or simply asking a question. Access is by password and the information falls into three categories: basic advice and descriptions, technique, and more detailed semi-academic and academic articles. While generic resources such as this have the advantage of being all-inclusive and very cheap on a per-head basis, it may in many cases be more appropriate to build a more limited resource that focuses very specifically on the particular program application. For example, a program aimed at young people at risk may need to focus much more specifically on issues such as child protection.
- *A monitoring and evaluation system.* There are two options here, which are not necessarily exclusive. The first involves recording each interaction, or at least each meeting, between mentor and mentee, or coach and coachee. You may simply want to monitor frequency of meeting, as an indicator of relationship health, or you may wish to capture in addition something about the themes explored in the virtual meetings. The latter raises issues of confidentiality. Electronic written correspondence is relatively easy to monitor, in both respects—basic logging systems can capture the former and the latter is simply a matter of monitoring for key words and phrases. It is impor-

tant to be upfront in how you intend to use this data: Is it to protect participants or to identify key themes that the organization needs to be aware of? (For example, content analysis of e-coaching in a European organization found that there was a much higher incidence of bullying behaviors in the workplace than formal records showed.)

The second option involves timed automated on-line questionnaires to participants. For example, Mentoring Dynamics System[2] examines a range of factors relating to goals, relationship experience, behaviours of mentor and mentee and outcomes for both parties. Three questionnaires are automatically sent out to participants, one each at the beginning of the relationship, one after six months and one after 12 months. The system issues reminders to non-respondents and informs the program coordinator who has and hasn't responded.

Useful questions to consider in this aspect of monitoring and evaluation include:

- Do we want to benchmark different programs or groups within the program against each other?
- Do we want to benchmark against other, external programs (for example, in other organizations)?
- How frequently do we want to measure?
- How will we use the measurement information when we have it?
- Will participants join at set points in the year, or will they join at random times? (If the latter, then people will be at different stages in their relationship, so aggregate data may be misleading, if it is gathered at set points in the year.)
- Who should have access to the monitoring data?
- In what form will reports be most useful to you? (Designing the output at an early stage is much cheaper than redesigning later!)

- *A training management system.* The difference in terms of success rates between programs where all participants are trained and those where training is not a pre-requisite is stark. A training management system enables the program coordinator to:

  - Ensure that all participants receive at least basic training before they start their relationship
  - Schedule training sessions to meet the needs of different groups (time, location, etc.)
  - Ensure that all participants receive timely joining instructions, reading materials, etc.
  - Offer advanced training, if available, to coaches/mentors who have had several coaching/mentoring relationships

The simplest systems are based on Excel spreadsheets, but others may include applications to book places on training events, issue invoices, if appropriate, and send automatic confirmations and reminders.

- *On-line training resources.* Many programs still insist on face-to-face training for participants, on the grounds that it is difficult to practice behavioral skills in any other way. Moreover, the majority of computer-based training in coaching and mentoring is not of good quality. However, it is quite feasible to design training approaches that involve a mixture of:

  - Pre-reading
  - CBT-based exercises
  - Reflection notes
  - Video-conference or e-mail-based practice and review sessions, moderated by an experienced on-line trainer.
  - Co-learning dialogue between participants
  - Demonstrations of coaching and mentoring techniques, either pre-recorded or live.

  The multi-media approach lends itself particularly to learning through supervision. In the context of virtual coaching and mentoring, the supervisor can review the on-line dialogue and/or telephone dialogue, offering observations and making suggestions about both process and content. In our view, this supervisory support is the most important aspect of the training package—yet it is the one least often provided.

- *The home page website.* The program website may have a wide variety of functions, including attracting potential participants to the program, keeping participants informed about program news, and providing the "first click" entry into all the other electronic functions. Key questions include:

  - How much of the website do we want to be transparent to all? To participants? To the program manager?
  - How integrated do we need the various functions we have just reviewed above to be (given that, for example, the training management system may be from a completely different provider than the on-line learning resources, or the measurement software)? To what extent do we need to import data from one function to another? Can we do so without great expense, or is it cheaper and easier to transfer data manually?
  - How do we keep the content up-to-date, relevant and of real interest to users?
  - What sort of back-up and restore procedures do we need to have in place?

IN SUMMARY

The costs of IT, both hardware and software, are falling. But the costs of badly designed systems are not. In designing a virtual coaching or mentoring program, investment of time at the beginning, defining precisely what technology is needed and how it will be supported, will pay dividends.

Many of the systems continue to be designed and developed without the full understanding of the needs of the users. It is imperative that participants' input is sought from the outset and is continued to be used during the design, development, testing, and implementation of the system. A more rigorous adherence to the Verification (are we building the product right?), Validation (are we building the right product?) and Testing (VV&T) process is needed.

Virtual coaching and mentoring not only helps overcome time, distance, and cultural barriers, but it also helps bridge the gap between the 'haves' and 'have nots,' thereby reducing the digital divide and increasing access to opportunities.

Technology is a great enabler of virtual coaching/mentoring. This helps meet the needs of participants in a flexible and meaningful way in a rapidly changing market place.

NOTES

1. http://www.clutterbuckassociates.co.uk/content/Company/Products/CAMeO.aspx
2. http://www.clutterbuckassociates.co.uk/content/Company/Products/MDS Online.aspx

REFERENCES

Garvey, B. (2004). When Mentoring Goes Wrong. In Clutterbuck, D., & Lane, G. (Eds.), *The Situational Mentor* (pp. 160–177). Aldershot, UK: Gower.

Hale, R.I. (2000a). The dynamics of mentoring as a route to individual and organisational learning. Doctor of Management Thesis, International Management Centres Association with Southern Cross University.

Hale, R.I. (2000b). To match or mismatch? The dynamics of mentoring as a route to personal and organisational learning. *Career Development International, 5*(4/5), 223–234.

Reiss, S. (2008). *The normal personality.* New York : Cambridge Press.

CHAPTER 4

PERSPECTIVES ON E-DEVELOPMENT[1]

Paul Stokes

ABSTRACT

Increasingly, we are using electronic media to make social connections between people. As well as the more established conventions of telephones and electronic mail, additional modes of communication are becoming available via electronic means—for example, the use of blogs, chat rooms, and other social networking facilities such as My Space and Facebook. All these have recently emerged to complement the use of personal websites. New applications such as Skype, for example, provide alternatives to conventional telephone calls, and integrative technologies such as Blackberries enable users to link email, text messages, and videoconferencing together.

In this chapter, I investigate the growing use of electronic media used to make social connections between people. This may include the use of email as well as dedicated coaching or mentoring software designed to facilitate developmental relationships. I take a brief look at the form the developing technologies are taking and consider some practical issues the impact of these innovations on coaching and mentoring.

Virtual Coach, Virtual Mentor, pages 77–84

INTRODUCTION

As Headlam-Wells, Gosland, and Craig (2006) point out in their work on e-mentoring, it is "a relatively new and under-researched field, particularly from a European perspective" (p. 273). As a result, the descriptions of e-coaching and e-mentoring are generally confined to talking about the use of email in an asynchronous manner to fulfill many of the functions of coaching and mentoring described thus far in this book. Indeed, our own research in this area (Megginson, Stokes, and Garrett-Harris, 2003; Megginson, Garrett-Harris, and Stokes, 2003) is essentially focused on schemes that used e-mail as the predominant mechanism for the delivery of mentoring. However, given the increase in the variety of media available, we need to make our starting point for e-development a little more inclusive. Initially, we will use the term e-development to refer to any coaching or mentoring process where the main mode of coaching and mentoring uses electronic means to connect people. This includes telementoring, videoconferencing, Skype, text, email, and other mechanisms that use the internet.

As is the case with many areas in mentoring and coaching, there is much more published research material that has come from the mentoring literature than the coaching literature. Indeed, the literature does not generally recognize the term "e-coaching." However, it seems reasonable to argue that many of the advantages and disadvantages of e-development apply equally to coaching initiatives as mentoring ones, particularly as much of discussion and comparison of such processes are between face-to-face interventions (common to both coaching and mentoring) and electronic ones.

In Megginson et al. (2006), Kate Kennett writes an account of her e-mentoring relationship with David Clutterbuck (216–219). Kate draws on this experience and identifies a range of issues for e-mentoring. One issue is as follows:

> All being delivered in *writing*, this process leaves a recorded trail of thoughts to which I can regularly return for further reflection. In this respect, I have found that E-mentoring has a definite advantage over face-to-face contact, for which an excellent memory may be required to recall an accurate account of a conversation. (Megginson et al., 2006, pp. 216–217)

Where transparency and accountability might be required—for example, in legal work, professional ethics, or in work with vulnerable adults or children—the benefit of a record of the content of a conversation can be particularly important. However, Kate also sets this against the potential loss of richness that a face-to-face developmental conversation may provide, she states:

> Virtual mentoring inevitably does not offer the wide range of communication and information that is available in face-to-face mentoring, depending as it

does pretty much solely on the written word. I think that this lack of opportunity to observe the mentor in action, 'read' his non verbal messages (and
he mine), and sense and hear complex intonation in the communication has
affected the potential richness of the mentoring relationship. (Megginson et
al, 2006, p. 218)

Kate is offering a different meaning to the idea of capturing or recording
the content of a conversation. Despite the arguable efficiency gains from using email for developmental '"conversations," the processes of summarizing
and paraphrasing what the learner says in a face-to-face conversation can be
particularly powerful. Karl Weick's (Weick, 1995) question: 'how do I know
what I think until I see what I've said' seems to resonate strongly with the
above point. Email, in particular, is less effective in capturing the richness
of a conversation and cannot convey much about the tone of voice or body
language. This presents both challenges and opportunities for the coach or
mentor who is working with the written word rather than the spoken.

Hamilton & Scandura (2003) offer an opportunity when they argue that
not having the social and visual clues can minimize the impact of "status
and social cues" such as gender, ethnicity, age, and other interpersonal factors that can influence the learner's perception of the help that they are
being given. They suggest that by removing these features, the mentee is
able to focus on the message rather than the messenger. Hamilton and
Scandura (2003) also argue that e-mentoring by email can help by making
the mentoring more accessible to more people (due to its ability to overcome geographical barriers and its asynchronous nature). Because of these
advantages, they argue that e-mentoring can increase the available pool of
mentors as well as their diversity (which may be crucial in some schemes).
In summary, they make the following claim:

> A mentor is a guide, role model, counsellor and friend. As long as these
> functions are being performed, the mentor's organizational location in rela-
> tion to the protégé is immaterial to the success of the E-mentoring relation-
> ship. . . . Whilst research on E-mentoring is just beginning, initial concerns
> regarding the lack of face-to face interaction and a decrease in the richness of
> communication may not be as much of an issue as initially assumed. (Hamil-
> ton & Scandura, 2003, p. 400)

This claim is supported by the recent work of Shpigelman, Reiter, and
Weiss (2008) in a small pilot study in Israel. This study was unusual as it
focused on participants with physical and mental disabilities being involved
in mentoring relationships. Although it was a very small-scale study, the preliminary results seemed to support the idea that effective helping relationships can be created online, with the impact of disabilities being minimized

due to the electronic mode of communication. The following case study also adds weight to this argument.

MENTORSBYNET—AN E-MENTORING PROGRAM FOR SMALL TO MEDIUM ENTERPRISE

Megginson, Stokes, and Garrett-Harris (2003) presents an evaluation of an e-mentoring project for the Small Business Service in the South East of England. This was a piece of evaluation work conducted by the author, David Megginson and, in particular, former colleague Ruth Garrett. The project's aim was to develop and grow the skills, knowledge, and confidence of SME owner-managers with a view to helping them succeed. The participants were entrepreneurs or small business managers. The pilot for this project lasted for over three months and was delivered entirely by electronic means. This included the support of a web-enabled online tutorial with four modules on mentoring and evaluation.

The research method used for the evaluation involved pre- and post-surveys. These surveys examined factors such as the perceived experience of the program, program outcome, satisfaction with contact frequency, and satisfaction with online training. These measures were similar to a comparable Australian study (abbreviated to APESMA); this Australian study was used as a benchmark for the UK evaluation. The key findings are summarised below:

- 96% of mentees and 80% of mentors described their e-mentoring experience positively. This compared favourably to the APESMA study, where only 82% of mentees described the experience as a positive one, while the mentor response was similar to APESMA at 80%.
- 91 % of mentees and 84 % of mentors indicated that they would participate in a similar program at some time in the future.
- Over 60% of mentees and over 70 % of mentors cited convenience, flexibility, and ease as the major benefits of email-based mentoring, while 30% of mentee and mentor responses indicated there is an element of impersonality about this type of communication.
- Over 50% of mentees and mentors indicated that they were planning on or thinking about continuing their relationship after the conclusion of the pilot.

This scheme touches on several of the emergent themes that have been identified in the opening remarks of this chapter. Firstly, as Hamilton & Scandura (2003) suggest, the experience of being mentored (irrespective of the mode used) seems to have resulted in a positive experience for those involved overall. However, as Kate Kennett (Megginson et al., 2006) suggested, the lack of face-to-face contact was an issue for at least a third of the participants. Other aspects of the feedback were also interesting. For example, there was a significant disparity between what participants expected to receive and what they actually felt they got out of the program. Prior to mentoring, 95% of mentors had expected to gain some personal development, while only 54 % of them felt that was what they had in fact received in the post-evaluation survey. Similarly, 90% of mentees stated before the program that they expected specific benefits in terms of improved business practices, whereas only 44 % of them felt that they received that.

We have to treat these findings with some caution because the pilot involved only 87 people with a 50% response rate for mentees and 59% for mentors. Also, there is no way of examining whether the participants' own assessment of their development is accurate. Nevertheless, as in face-to-face mentoring and coaching, these findings do draw our attention to the importance of contracting and development processes in terms of setting realistic expectations of what the mentoring can or cannot achieve. Although this was included in the online tutorials, this raises some questions about whether it is possible to engage with mentoring skills development without some aspect of experiential learning.

DISCUSSION

In many ways, the findings from these case studies are unsurprising. Putting aside the contribution made by the book within which this chapter sits, it seems obvious to argue that electronic mentoring and coaching has been under-researched thus far. This lack of empirical work underlies the wider challenge to researchers within the coaching world specifically to develop more academically robust accounts of the work done on coaching. Mentoring (see Garvey, Stokes, & Megginson, 2009, Chapter 2), however, has a different and richer research tradition so it might be expected that one would find a richer vein of empirical work on e-development within this literature. While this is certainly the case relative to coaching, it still seems fair to argue that the protocols, timings, and efficacy of certain methods within e-development are not well understood as yet. With this in mind, I would like to offer a broad framework for categorizing e-development schemes and then expand on some of the challenges facing our understanding of these approaches.

THREE BROAD PERSPECTIVES ON E-DEVELOPMENT

Based on my research experience and reading of the field so far, I propose that there are three main approaches to using electronic media as part of coaching and mentoring. These are:

1. Pure E-Development—where all aspects of the coaching/mentoring are done using electronic means including recruitment, selection, development, matching, conversation, support, and evaluation
2. Primary E-Development—where the majority of the coaching and mentoring activity is done using electronic media but interspersed with some face-to-face meetings
3. Supplementary E-Development—where the use of electronic media for coaching and mentoring activity is seen as a useful add-on or additional aspect of the process but is not central to the scheme or process.

It is my view that most e-development falls into the supplementary e-development category. It is also my view that the majority of all coaching and mentoring relationships—be they formal or informal—are in the supplementary e-development category. This is simply because electronic media have become a central part of many peoples' lives. Inevitably, whatever the starting point for the relationship, participants in coaching and mentoring relationships will tend to create and develop ways of interacting with each other using a variety of methods. Although the core interactions—the substantive aspect of the relationship—may be face-to-face meetings, it is increasingly likely that these things will be followed up and supplemented with phone calls, emails, text messages, and other ways of maintaining and developing the relationship. Similarly, if the starting point is primary e-development, unless there are significant temporal, geographical, or other boundaries that prevent it, it is likely that face-to-face meetings will become part of the relationship at some point, even if the relationship remains a primary e-developmental one overall.

In her book *In the Age of the Smart Machine,* written more than 20 years ago now, Shosana Zuboff (1988) discussed how information and collaborative technologies would have the potential to radically transform communication and knowledge within organizations. She posited both a positive and negative version of this for organizational life in terms of the degree to which it would both enable and constrain the people who used it. Since then, writers such as Handy (1994), Sennett (1998), and Beck (2000) have continued this ambivalence towards the future and the impact of the use of technology on the world of work, recognizing some of the opportunities for some people but the possible sense of exclusion for others. In similar vein, further investigation is required to establish the extent to which e-development, in its various forms, might add to as well as take away from what coaching and mentoring can offer individuals, groups, and organizations. More practically, despite there now being a plethora of media for communication, there is, at present, relatively little understanding of how such mechanisms work in the context of a coaching and mentoring conversation. What are the appropriate protocols, for example, to be used between an executive coach and a senior manager, where the latter relies heavily on her blackberry as their primary means of communication? If my theory of "blended" virtual and face-to-face conversations is correct, this then begs the question, what is best blend? As with many things in coaching and mentoring, the answer to this question is likely to be driven around (a) individual preference and (b) context. However, it will nevertheless be important to take into account the ubiquity of mobile phones, blackberries, laptops, webcams, etc., all of which have the potential to contribute to coaching and mentoring conversations. In order to take these things into account, we as enquirers into the world of coaching and mentoring must be prepared to refine our research methodology to capture a sense of what media are actually being used (as opposed to what 'should' be being used) between coach/mentor and client. How often, for example, will a client text his coach/mentor as opposed to calling her on her mobile phone? How often do participants use the electronic resources like the chat room facilities or blogs? What uses, and to what extent, do people tend to put them to? Following that, it will then be important to understand how these different media might be used and what the promises/pitfalls of each might be for coaching and mentoring conversations. For example, to what extent do the use of email and telephone coaching sessions serve to perpetuate a coaching/mentoring relationship? If it does, is this always useful or might it militate against a '"good ending"? Does regular "surface" contact by text message serve to "crowd out" more substantive face-to-face contact, or does it add value?

In summary, therefore, there is much that is not yet understood about the blend and interaction of the modes of coaching and mentoring and the impact they have overall on the relationship. While, judging by some studies (eg Headlem-Wells, Gosland, and Craig, 2006), we are getting better at understanding how to deliver e-development programs, there is still room for increased focus on the mix of methods by which such outcomes are achieved.

NOTE

1. An earlier and adapted version of this chapter is published in Garvey, Stokes & Megginson (2009). *Coaching & Mentoring Theory & Practice*. London: Sage.

REFERENCES

Beck, U. (2000) *The Brave New World of Work.* Cambridge: Polity Press.

Garvey, B., Stokes, P., & Megginson, D. (2009). *Coaching & Mentoring Theory & Practice.* London: Sage.

Hamilton, B.A., & Scandura, T.A. (2003). Implications for Organizational Learning and Development in a Wired World. *Organizational Dynamics, 31*(4), 388–402.

Handy, C. (1994). *The Empty Raincoat: Making Sense of the Future.* London: Hutchinson.

Headlam-Wells, J., Gosland, J., & Craig, J. (2006). Beyond the Organisation: The Design and Management of E-Mentoring Systems, *International Journal of Information Management, 26,* 272–285.

Megginson, D., Stokes, P., & Garrett-Harris, R. (2003). MentorsByNet—an E-mentoring Programme for Small to Medium Enterprise (SME) Entrepreneurs. Evaluation Report on behalf of Business Link Surrey, UK.

Megginson, D., Garrett-Harris, R and Stokes, P (2003) Business Link for London E-mentoring scheme conducted by Prevista.biz—for Small to Medium Enterprise (SME) Entrepreneurs/Managers, Evaluation Report for Business Link London.

Megginson, D., Clutterbuck, D., Garvey, B., Stokes, P., & Garrett-Harris, R. (2006). *Mentoring in Action: A Practical Guide.* London: Kogan Page.

Sennett, R. (1998). *The Corrosion of Character: The Personal Consequences of Work in the New Capitalism.* New York: Norton.

Shpigelman, C., Reiter, S., & Weiss, P.L. (2008). Ementoring for youth with special needs: Preliminary results. *CyberPsychology & Behaviour, 11*(2), 196–200.

Weick, K. (1995). *Sensemaking in organizations.* London: Sage.

Zuboff, S. (1988). *In the Age of the Smart Machine: The Future of Work and Power.* New York: Basic Books.

CHAPTER 5

CRITICAL SUCCESS FACTORS IN E-MENTORING FOR SMALL BUSINESS

Dr. Kim Rickard

ABSTRACT

While information communications technology provides new opportunities for supporting mentoring, there is a need to explore the determinants of effectiveness in this new environment. Using qualitative enquiry, the analysis establishes patterns in the characteristics of effective and ineffective e-mentoring partnerships, which in turn forms the basis for a set of critical success factors for use by those developing and evaluating the effectiveness of e-mentoring programs.

This study is an abridged version of a broader effectiveness evaluation undertaken that included a detailed program description, a quantitative analysis of effectiveness, an analysis of the implications of the constructivist paradigm stance adopted, extensive qualitative data rather than the indicative quotational data included here, and analysis of anomalies that arose in the data. The quantitative analysis was conducted using a survey questionnaire that provided the basis for classifying the sample used in this analysis.

Virtual Coach, Virtual Mentor, pages 85–108
Copyright © 2010 by Information Age Publishing

INTRODUCTION

Structured e-mentoring is defined as a partnership between a mentee and mentor using email as the primary mode of communication within a formalized program environment that provides matching, training, and content around which mentee and mentor engage.

The predominance of speculatively based e-mentoring studies, and the lag between practice and research (Perren, 2002) has meant that e-mentoring program development and evaluation may not be founded on an empirically based understanding of the determinants of effectiveness.

Using a model derived from DeLone and McLean's (1992, 2002) model of Information Systems Success and respecified for the structured e-mentoring environment (Rickard, 2007), this study establishes a range of factors linked to effectiveness arising from an examination of actual practice in the small business context. Approaching the evaluation of e-mentoring with reference to a multi-dimensional systems approach is useful because e-mentoring has the characteristics of complexity with a multiplicity of variables, a high degree of influence between the various dimensions and the context into which it is placed, and the fact that it is a complex construct best evaluated when all its parts are considered not in isolation but in relation to its constituent dimensions. Effectiveness will be evaluated in terms of (a) the nature and quality of the mentoring partnership, (b) the nature and quality of the e-mentoring program structure and content, (c) program use, (d) user satisfaction, and (e) impact. The model to be used as a basis for the classification, description, and interpretation of the quotational data set out in this chapter is summarized in Figure 5.1.

A total of 20 mentees participated in this study, and the interpretations are based on extracts from email exchanges, responses to open questions in the survey questionnaire, and semi-structured mentee interviews.

In this study, the effectiveness construct was operationalized as shown in Table 5.1.

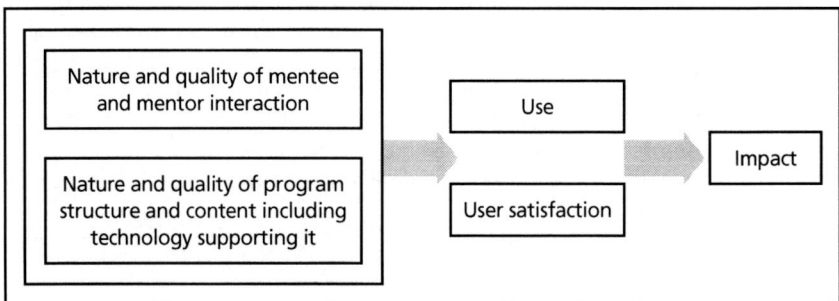

Figure 5.1 Rickard model of structured e-mentoring effectiveness as derived from DeLone and McLean's 1992 and 2002 model of IS success.

TABLE 5.1 Operationalization of Effectiveness Construct

Dimension of e-mentoring effectiveness	Measures selected to operationalize
(i) Nature and quality of mentoring relationship	• Nature and quality of mentoring relationship • Mentor as impartial or neutral sounding board • Types of advice and support provided

(ii) Nature and quality of program content and structure	• Quality of match
	• Nature and quality of program structure
	• Development of individualised learning pathways
	  – Adaptation of program structure and content
	  – Personal goal setting
	  – Integration of program with business activities

(iii) Use	• Interaction frequency • Impact of email delivery on use
(iv) User satisfaction	• Mentee's perception of value
(v) Impact	• Learning by mentees • Benefits—including long-term • The degree to which mentees' needs were met • Unexpected as well as anticipated outcomes

FINDINGS AND DISCUSSION

The Nature and Quality of the Relationship between the Mentor and Mentee (Table 5.2)

The literature suggests that this is the critical dimension to which effectiveness is linked (Devins & Gold, 2000). As set out in Table 5.1, this dimension is operationalized with reference to the nature and quality of the mentoring relationship, the mentor as sounding board, and the types of mentee support sought and received.

Mentees involved in effective e-mentoring partnerships described their relationship with their mentors as being invaluable, important, a strong bond, fundamental to positive outcomes, open and honest, involving mutual respect, and the mentor as identifying with their needs and comfortable to communicate with.

In contrast, mentees involved in ineffective partnerships described their mentor as not interested in listening or understanding, disregarding the needs and wants of the mentee, as having their own agenda, and the mentoring relationship as being limited in scope. The comparative analysis of the qualitative data for effective and ineffective partnerships supports the proposition that there is a strong relationship between effectiveness and the nature and quality of the mentoring relationship.

TABLE 5.2   Nature and Quality of the Mentoring Relationship

Effective

- The quality of my relationship with my mentor was fundamental to the positive outcomes because she was very quick to respond to my questions, gave me her personal insights and experiences and supported these with published articles on the area which we were reviewing. Also every question and personal observation I made was addressed thoroughly and with a sense of warmth so I felt she identified with my thoughts and issues, which gave me confidence in the relationship. Her answers were simple and straight to the point and she always knew exactly what I was asking, so her answers were always relevant. The articles she attached to her responses provided more depth and I could save them and read them several times. This gave me a lot of confidence so that I could move forward.
- My partner and I were geographically separated and indeed we only contacted by email. So related to that, how much of a quality relationship can you generate in cyberspace? Actually, I think you can do a lot. I run several companies with a partner who is 1500km away, and have done so successfully for 4 years. Plus I have long term friends that I developed in cyberspace . . . I think that a quality relationship can be generated and it is vital—to a successful program.

Ineffective

- From the initial telephone conversations it was apparent . . . that the Mentor was keen on talking, but not too hot on listening. [The] . . . relationship would have been strengthened if [my] mentor was able to listen more and in particular, sought to understand in detail the issues we were facing.
- Mentor . . . provided "canned" advice.

Mentor as Sounding Board (Table 5.3)

The literature suggests that impartiality is considered one of the major advantages of e-mentoring, and on this basis, the following section will explore whether or not access to a neutral and impartial sounding board may be related to effectiveness.

The data suggest that effective mentoring partnerships often involve the use of the mentor by the mentee as a sounding board. The mentees involved in effective partnerships describe their experience as involving the use of their mentor for a different perspective on their business, to bounce ideas, to lift their focus or to see things differently. In contrast, the responses of mentees involved in ineffective partnerships describe their mentor as not listening or as interested only in one-way information flow.

Types of Support (Table 5.4)

The literature confirms that diversity of support and advice is a characteristic of effective mentoring (O'Neill, 1998; Clutterbuck, 2003). This

---

TABLE 5.3   Mentor as Sounding Board

Effective

- It's funny how someone else can see things just that little bit differently and can start you thinking again!
- It was fantastic to have someone to bounce ideas off who has been there before.
- There truly needed to be a third party...to force the individual to reappraise the big picture.

Ineffective

- I was looking for a sounding board to share ideas [but the] mentor appeared to be only interested in one-way information flow.

TABLE 5.4   Diversity of Support and Advice Sought or Provided by Mentee

Effective

- I felt that...I could ask my mentor about anything, not just about work things but also about work-life issues and how she managed.
- [My mentor] was helping me to get a clear picture and ideas which I needed at the time...Effective mentoring to me is when the mentor is able to encourage the mentee to tackle other angles of problems and look at other options. The mentor should not be offering solutions, but should be helping the mentee to explore the options and see other solutions themselves. The mentor should be opening up the avenues of thought and at the same time being able to give practical advice on standard business issues such as suggested effective processes on how to lay out a business plan and conduct a SWOT analysis, etc. I believe we achieved this.

Ineffective/partly effective

- I had a specific set of issues....
- [The program was] helpful for answering specific questions.

---

section will consider whether there is evidence to this effect in the context of structured e-mentoring, and then whether or not it is possible to draw a credible link between the diversity of support and advice provided to mentees and effectiveness.

There was strong evidence to suggest that effective relationships were characterised by diversity, breadth, and depth in advice and support/s provided, while a degree of specificity in mentee expectations and support sought characterized the ineffective partnerships. These data suggest a link between effectiveness and wide-ranging support sought or received but do not clarify the ambiguity regarding causal direction. The lack of advice sought in wide-ranging areas in the ineffective partnerships could be either an antecedent to effectiveness or it could be the outcome of an ineffective partnership.

THE NATURE, QUALITY AND ADAPTATION OF PROGRAM STRUCTURE AND CONTENT

The literature suggests that this dimension is critically linked to effectiveness. Boyle and Boice (1998; cited in Boyle-Single & Muller, 2001) suggest that structured e-mentoring practitioners have attempted to address many of the problems related to the unstructured nature of many mentoring programs, and that provision of such structure is positively linked to effectiveness: "Proper program structure and personnel improve participant involvement and increase the benefits associated with mentoring programs" (p. 109).

As set out in Table 5.1, this dimension is operationalized with reference to quality of match, the nature and quality of program structure, and development of individualized learning pathways.

Quality of Match (Table 5.5)

Mentees involved in effective partnerships described the quality of the match in terms of a shared journey, having travelled the same path, having a similar outlook or character, being on the same track and there being a synchronicity of personalities. In contrast, mentees involved in ineffective partnerships described the match in terms which indicated less or no such similarity or synchronicity.

TABLE 5.5 Quality of Match

Effective

- We were so well-matched because she had a family and had needed to travel extensively, so immediately understood the demands, and I never had to overly explain my lack of contact or delayed responses. If I mentioned some of the anxieties I had about travelling, demanding schedules and leaving one's children, she always had a similar story and injected a sense of humor in the telling, which eased my stress.
- I felt the existence of a shared journey even though our areas of expertise are 180 degrees apart. My mentor had travelled the same paths as me in terms of family, demanding schedule, lots of overseas travel, working as an employee when my skill set would be more suited to running my own company.
- The feeling of relevance of the mentoring partner's work/life experience to one's own and the synchronicity of the personalities involved . . . affected [the outcomes].

Ineffective/partly effective

- I am . . . extremely disappointed with the . . . matching process; I communicated that I was willing to offer my services/advice to somebody that was passionate about developing their business or at least was struggling and needed help.
- Was bad match at interpersonal level and had major impact. Skill set and background of mentor was reasonable match.

Nature and Quality of Program Structure (Table 5.6)

The data indicate that mentees involved in both effective and ineffective partnerships found the structure useful in similar ways, but also shared some concerns about the format.

While overall the email facilitation messages operated as intended, there was some evidence to support the proposition that facilitation messages "dragged down" mentees or made them feel that they weren't completing the program as expected if they didn't address all of the issues set out in the email messages. The data indicate that the email facilitation messages in some cases may have negatively impacted on the effectiveness of the program for mentees.

Comparison of Divergent Experiences—Effective and Ineffective

In comparing the quotational data for mentees involved in effective and ineffective mentoring partnerships, there is some evidence to suggest a link between effectiveness and mentees' positive experience of the program content and structure.

Mentees generally indicated that they found the program useful in that it provided a structure with timelines/deliverables and prompts for action, and helped manage expectations, provided a framework, assisted with identifying possible areas on which to work, maintained momentum, focused

---

TABLE 5.6 Nature and Quality of Program Structure

Effective

- It was great to have a structure within which to work—that included timelines/ deliverables, because this sort of structured program so rarely happens.
- The . . . program structure was critical in the beginning as I had no idea of what to expect or how I should approach the role of mentee. It gave me guidelines on what to expect—I had not expected the amount of support and material provided and was anticipating having to do a lot more research to find answers.
- It provided a framework for us and also assisted us in identifying possible areas on which to work and clarifying what my needs were and what wasn't relevant.
- It is often a scary thing going into this and the email support and info was invaluable.

Ineffective

- It was a good start to make sure we started focusing on key elements.
- I found the messages too wordy. I simply didn't have time to fully read and digest them. Shorter, more succinct messages would have been better. Also the messages seemed to simply add more load. They made me feel a bit guilty that my mentor relationship was not progressing fast enough—that I wasn't quite up to the task.

on the finite amount of time available, provoked discussion, and defined where to begin and end. In these ways, the program operated as intended in accordance with "stated ideals." In contrast, the quotational data suggest that those mentees involved in ineffective partnerships may have disproportionately experienced difficulty with the length and "wordiness" of the email facilitation messages.

The comparison of quotational data indicates that a mentee's response to the program content and structure may not only be critical to facilitating effectiveness, but may also be a factor in program ineffectiveness.

Construction of Individualised Learning Pathways

The program structure and content was intended to assist participants with constructing their own learning pathways in the form of adaptation of content, goal-setting, and integration of the program with the mentees' business activities.

Adaptation (Table 5.7)

Participants in the program were encouraged to adapt the program to meet their own needs and to disregard exercises or activities which were not relevant to them.

TABLE 5.7   Adaptation of Program Content and Structure

Effective
- We selected the parts that were relevant to my situation at the time and ear-marked others for the future.
- We did not follow the classical format. In part due to geography but also personalities.
- After some time I said that we should abandon the "program's structure" and just communicate and focus on "low hanging fruit," getting some achievable goals under our belt. That, combined with a conference that the mentee went to established a clear picture of strategy. With a renewed focus the mentee chose to look at achievable goals, simplify everything, reduce risk and anxiety. This was an epiphany for him and he was very thankful for it. He is now in a much happier place and delighted with the outcome. So simply put there was a lot of stress before hand, and the programs nature added time constraints and pressure. By abandoning the "compulsory" nature of the steps so as to not appear behind—we developed a freedom to think and prioritize.

Ineffective
- [The mentor's] ... input seemed to be very "potted" and not well tied to the actual conditions/issues facing me and my company.
- ... the mentor wanted to follow his pre-conceived plan of action so things went from bad to worse.

The qualitative data suggest a strong link between effectiveness and adaptation of program structure and content. The mentees involved in effective mentoring partnerships reported that they were selective in their use of content, used only relevant content, followed some parts and not others, picked essential elements, found the right blend, changed the timing and regularity of communications to suit their needs, and in some cases abandoned the structure altogether.

In contrast, mentees involved in ineffective partnerships generally reported that they did not adapt the program to the extent evident in the quotational data arising from effective partnerships. These mentees described their mentor's input as "potted," and the mentor as not adaptable, having a pre-conceived plan, and mentoring partners as not adapting to each others' learning styles.

The quotational data suggest a positive link between effectiveness and whether participants adapted the program structure to their own needs. There is evidence to suggest that this process of active adaptation is useful and likely to contribute to maximizing effectiveness.

McLaughlin (1976; cited in Patton, 1990 p. 106) says:

> Where [program] implementation [is] successful, and where significant change in participant attitudes, skills and behaviour occur[s], implementation [is] characterised by a process of mutual adaptation in which project goals and methods [are] modified to suit the needs and interests of the local staff and in which the staff changed to meet the requirements of the project. (p. 106)

The link between effectiveness and the extent to which mentees and mentors changed or adapted the generic content and activities provided by the program host confirms McLaughlin's view and suggests that an analysis of the adaptation process is critical to understanding the effectiveness of structured e-mentoring.

Personal Goal Setting (Table 5.8)

Participants in the e-mentoring program were encouraged to set program goals to work towards throughout the program.

The quotational data set out in Table 5.8 indicate a link between goal-setting and effectiveness. Mentees involved in effective partnerships generally reported developing a range of goals. The goals were not always formalized, well-defined, or maintained, but mentees generally suggested that they were important to getting the most out of the program. In contrast, the ineffective partnerships were fairly clearly characterized by the failure to set goals.

TABLE 5.8   Goal-Setting

Effective

- I had no "goal" list, but through the program my mentor may have picked up on areas that needed her attention.
- [Developing a range of goals was] ...a critical function, and then with a lack of progress we redefined the goals and got back on track.
- I am sure that many enter the program unclear as to what the goals are. It is sort of well, maybe something good will happen if not what have I lost. Trust comes with time. Trust leads to motivation and execution so it does improve with time as the relationship develops.

Ineffective

- We didn't [set goals] but if relationship continued this would have been good.
- No [we didn't set goals].

Integration (Table 5.9)

The program was based on a situated learning model and encouraged mentees to integrate the mentoring process with their day-to-day business activities. In Hartshorn and Parvin's (1999) terms, the program took a "naturalistic" approach which drew on this situated learning theory.

The comparative analysis indicates that mentees involved in effective partnerships integrated the program with their business activities. These mentees reported integrating their mentor's suggestions into daily activities, considering critical incidents in terms of daily business activities, and getting help from the mentor on day-to-day issues. In contrast, mentees involved in ineffective partnerships only rarely asked questions directly related to day-to-day business activities, and fitted mentoring around rather than making it part of their business activities. The quotational data provided some initial support for making a link between integration of the mentoring program with business activities and effectiveness.

Use (Table 5.10)

The dimension of use is generally defined as involvement and considered with reference to the frequency and duration of e-mentoring interactions.

Analysis of the data set out in Table 5.10 suggests a link between effectiveness and reported interaction frequency, with all those involved in effective partnerships acknowledging that regular contact was important. Many of those involved in effective mentoring partnerships, however, qualified their statements about the importance of regular interaction with comments about the importance of the quality of interactions as well as their frequency.

TABLE 5.9   Integration of Program with Business Activities

Effective

- During the day I would jot down issues and thoughts to discuss, but communication with my mentor did not happen every day.
- Highly integrated because it mostly used email to communicate which I use daily.
- I am still integrating her suggestions into my daily activities.

Ineffective

- Didn't get special time, just had to fit in around work.
- Work commitments were often main priority over mentoring.

TABLE 5.10   Use Interaction Frequency

Effective

- I was surprised at how emotionally attached I got to the program and would really look forward to emails. So I guess I found regular contact really important.
- The frequency was not as important as the fact that her replies were so rapid and the matching of her responses in terms of length, amount of information and relevancy of the information was fundamental to the success.

Ineffective

- Dialogue was too slow—2 days turnaround.
- We never established a schedule for regular communication . . . The communication we did have was very valuable though.
- [I] was unable to elicit commitment to regular communication from my mentees.

The quotational data, therefore, indicated the importance of considering the frequency of exchanges alongside their content and quality.

Email Program Delivery (Table 5.11)

The convenience and flexibility of email communication was acknowledged by those involved in effective and ineffective partnerships alike. The advantages seen by participants were fast turnaround on messages; the avoidance of polite chat and focus on the main issues; the provision to reflect on, review, and consider suggestions and responses; the tendency for the process of writing responses to clarify the issues for the mentee/writer; the capacity to make use of the asynchronous nature of email communication to juggle other commitments; and the cover of anonymity providing a less threatening or confrontational way of raising and discussing issues.

The suitability for those with busy business schedules and the decreased likelihood of participating in any form of mentoring program if it were not email-based is widely commented on by interviewees involved in both

TABLE 5.11   Email Program Delivery

Effective

- Face-to-face would be much harder to fit in because of distance apart and time commitments, whereas e-mentoring is always available, and because it is written it helped me enormously in clarifying the problems and assistance I required. It also allows the sending of extracts, printed notes and published material that I would have had to source or my mentor print and hand to me. Also e-mentoring allows you to write down and clarify the issues immediately as they arise, so you don't forget things. My visual memory is much better than aural, so it really worked well for me. While face-to-face mentoring would have advantages, the time and geographical limitations mean it would be much less likely to happen and to maintain the meetings. Also I think visual communication cues are more complicated and if both mentor and mentee can express themselves well in the written form, there is no loss of meaning or feelings. E-mentoring also allows a written record of the communications, which can be revisited at any time.
- I am really comfortable with email, and in some ways it probably meant I could be more frank because I didn't have to face my mentor. Face-to-face mentoring would be different, but not necessarily better.
- Flexibility regarding timing and the opportunity to consider and carefully weigh issues and responses/questions before sending them.

Ineffective

- Would have been better if mentor was available . . . for face-to-face meetings I suspect.
- Response time could be fitted in with other work commitments. Email provides written record of interaction.

effective and ineffective partnerships. As Single and Single (2005) suggest, "E-mentoring practitioners and researchers have not suggested that e-mentoring replace face-to-face mentoring, but have viewed it as a way of providing mentoring opportunities that otherwise would not exist" (p. 305). The fact that difficulties with the e-mentoring format were common to mentees involved in both effective and ineffective partnerships suggests that not only will e-mentoring not replace face-to-face mentoring, but that the e-mentoring format can impact negatively on those involved in both effective and ineffective partnerships. E-mentoring may in fact not be suitable for some even when other mentoring options do not exist.

Similarly, mentees involved in both effective and ineffective partnerships detailed the disadvantages of text-based computer-mediated communication. Email was described as less appropriate in some circumstances, and as a less efficient form of mentoring than face-to-face.

The quotational data suggest that those involved in effective and ineffective partnerships did not experience the advantages and disadvantages of email-based communication differentially. It is acknowledged that the data collected and presented in this section is limited. What it did establish, however, was that the choice of email as the technology underpinning this structured e-mentoring program was a facilitative factor for many participants involved in both effective and ineffective partnerships. A link between effectiveness and the use of email as a positive support for the mentoring process was established in these terms.

User Satisfaction

The dimension of user satisfaction is defined with reference to mentee perceptions of value. The literature refers to user satisfaction as a relevant and appropriate, though not always reliable or sufficient, measure of effectiveness.

Table 5.12 sets out quotational data that provides an indication of the level of user satisfaction by participants. The data are ranked in order of effectiveness scores from most to least effective as found by the quantitative analysis (refer to chapter summary/abstract).

TABLE 5.12   Matrix of Quotational Data Indicating Level of User Satisfaction

Participant ranking	Quotational data indicating level of user satisfaction
Effective	
1	Was great to be in contact with [my mentor]. Receiving feedback and advice from an experienced industry professional [was the most valuable part of the program].
2	This is one of the best things I've ever ever done . . . a thousand thank-yous. It really was magic to be able to share such similar experiences.
3	I've got SO MUCH out of the program . . . I have grown so much both professionally and personally through my participation in the . . . mentoring program. I found the experience really enlightening. It really helped me to fast-track lots of aspects of my business.
4	I loved it. I am having such a good time I do not want it to end.
5	Basically I got affirmation that I was on the right track and that was important.
6	The scheme is an excellent one and I hope it continues. I feel we are all richer for the experience and [my mentor] has been a delight to work with . . . we hope to maintain some form of contact even if it is coffee on a fly by!
7	I found the program to be very helpful . . . it was a good experience . . . and it was good to have a "sounding board."
8	I have developed a good relationship with [my mentor] . . . I think we will each get something out of this in the long run. I have given the first draft of my business plan to [my mentor] for his comments . . . I look forward to continuing interactions with him.
9	The main problem that caused me not to get the best . . . out of the program was the fact that both myself and my mentor became very busy.

(continued)

TABLE 5.12 Matrix of Quotational Data Indicating Level of User Satisfaction (continued)

Participant ranking	Quotational data indicating level of user satisfaction
10	It gave me a sense that I could control the direction of my business.
11	I thank my mentor for his appreciation of the need for balance and that fact that we could both say we needed time out for a few days to reconnect with ourselves and our families whilst remaining aware of the need for direction and focus.
Ineffective	
12	[My mentor's] input has been invaluable.
13	…the mentoring relationship did not work particularly well.
14	…Just as you get into the swing of things, the program is over. I struggled a bit because there seemed to be so much to cover.
15	[Communication was limited but] [t]he communication we did have was very valuable.
16	I work full-time and…found it hard to commit the time needed to the program.
17	Didn't know how to fit into business—where to start—what problem to address (open question, questionnaire).
18	Good information was provided. Unfortunately I didn't make…use of it due to hectic time.
19	[The most valuable part of the mentoring experience was] [t]he mentor [herself] and her generosity… [The least valuable part of the mentoring experience was] emails from [the host].
20	Waste of time and effort.

The quotational data suggest that mentee descriptions of their level of satisfaction become progressively less satisfied in line with declining effectiveness scores. This confirms the proposed link between user satisfaction and effectiveness.

Impact

The dimension of impact is defined for the purposes of this evaluation as the benefits or outcomes arising out of the structured e-mentoring program.

Evidence of Learning (Table 5.13)

A comparison of the quotational data for mentees involved in effective and ineffective partnerships provided initial support for the proposition that there was a link between effectiveness and positive learning outcomes.

TABLE 5.13   Learning by Mentees

Effective

- Mentoring should ideally allow the mentee to make quantum leaps re: insights and options available and assist them in seeing alternatives. It should also tackle areas in which the mentee may need more confidence. [The program] prompted me to become more business-like and professional and to value my contribution more highly. Practical, down-to-earth solutions suddenly seemed clearer and I have changed to more productive methods.
- My commitment increased as the program progressed because when I felt the benefits I wanted the program to have a high priority instead of something that fitted in around everything else.
- I believe the benefits are enormous. For people who work alone and have no formal business training, it is an efficient, rewarding, and very satisfying way to obtain support and receive encouragement.

Ineffective

- I was looking for [a] sounding board to share ideas [but the] mentor appeared to be only interested in one-way information flow.
- No [I did not find the combination of structured content, mentor support and contact with the host useful as a framework for learning].

The mentees involved in effective mentoring partnerships described their learning in terms of seeing alternatives, providing insight, clarifying solutions to problems, and a way of developing new skills, knowledge and an enhanced mindset.

In contrast, those involved in ineffective partnerships described the type of learning offered as different from what was expected or desired, that advice was one-way rather than collaborative, and that the learning was limited in nature such as answering specific questions. In two cases of ineffective partnerships, the mentees indicated that they did not find the combination of structured content, mentor support, and contact with the host useful as a framework for learning, and two mentees indicated that the mentor and mentee did not adapt to each other's learning styles.

This contrasting data would suggest that the presence or absence of positive learning outcomes was fundamentally linked to program effectiveness for mentees.

Benefits (Table 5.14)

A comparative analysis of qualitative data provides support for the link between benefits and effectiveness. The quotational evidence described wide-ranging benefits including increased focus; business planning; a better understanding of business dynamics; personal growth; knowledge of

---

TABLE 5.14   Benefits/Outcomes

Effective

- This mentoring program has really got me focused again and got me thinking about what I want to achieve, when I want to and will be able to achieve these things. It's made me realize that the planning of my business can't be done in isolation from other aspects of my life.
- The benefits for me were mainly in having a sounding board. I also benefited from my mentor's business experience when it came to our business plan in particular. Having said that, the business plan was never completed and now we are starting to work with an external company to try and put together a one-page business plan. However, the work done on the original plan was useful and is going to contribute to our new plan.
- The program was extremely effective for me. I found some of the classic, generally male bullying behavior distressing, and one of the main insights I gained from the program was that it is a game. This understanding allowed me to field the shots and enter negotiations without feeling intimidated.
- [The major benefits were] personal growth, reminders of business best practices, problem-solving skills, relationship skills.
- Another huge benefit for me was that I know longer felt isolated, which is the downside of working from home and on one's own.
- [The major benefits were] understanding how business operates and the financial aspects. Being more aware of the market and sales process as well. [Also] marketing, business planning and sales.

Ineffective

- I cannot help you if you are not interested in developing a business and in particular if you are not interested in a mentor relationship. Based on the assessment of the situation, I am extremely disappointed and consider the process to date, a waste of time.
- [There were] NIL [benefits].

---

business best practice; improved problem-solving skills; relationship skills; reduced feelings of isolation; lateral thinking; renewed energy and persistence; improved strategic planning skills; affirmation; better time management skills; problem-solving skills; improved self-confidence; and greater awareness of business, finance, marketing and sales. In contrast, those involved in ineffective partnerships described the mentoring program as being a waste of time with no benefits, confirming the link between lack of benefits and ineffectiveness.

The quotational data support the view that, in the case of mentees involved in effective partnerships, there was considerable evidence of benefits, business skills enhancement, and achievement of anticipated outcomes.

Long-Term Benefits (Table 5.15)

While the data that provided the basis for the comparative analysis were limited, they nonetheless provided sufficient indicators of long-term ben-

TABLE 5.15   Long-Term Benefits

Effective

- Thanks for all the info so far. I think it will keep me going far beyond this mentoring program.
- Thank you so much for initiating the mentor program and for your guidance and assistance with it—it has changed my life in many ways. . . .
- Long term . . . my perception of my abilities changed and expanded and I felt myself changing in my mental approach . . . I now see more opportunities and feel a lot more pro-active and excited about the options for my career.

Ineffective

- No [no long-term benefits]
- No [participation in the program did not contribute to my business's long-term stability, viability or growth]

efit for those mentees involved in effective and ineffective partnerships to make an initial link between effectiveness and long-term benefit. Whereas mentees involved in effective partnerships described benefits as extending beyond the program, changing their life, providing long-term change to abilities and approach, and providing a sense of control in the long-term, mentees involved in ineffective mentoring partnerships indicated limited or no long-term benefits.

Meeting Mentees' Needs (Table 5.16)

A comparison of the quotational data for mentees involved in effective and ineffective partnerships suggests a link between effectiveness and meeting the needs of mentees. The mentees involved in effective mentoring partnerships described the ways in which their needs were met in varied but very positive terms such as addressing issues which were being avoided, developing new strategies to deal with problems identified, finding solutions to difficulties, breaking entrenched patterns of thinking, and seeing new opportunities. In contrast, mentees involved in ineffective partnerships described the mentor as not listening, not seeking to understand, not addressing their issues, and wasting their time and effort. This contrasting data would suggest that meeting the mentees' needs was fundamentally linked to effectiveness.

Unanticipated Outcomes (Table 5.17)

Scriven (1993) suggests that side effects should be sought and evaluated as "serious or trivial, fatal or fortunate" (p. 24). He points out that if a pro-

---

TABLE 5.16 Meeting Mentees' Needs

Effective

- The last few months of discussions with you have made a huge impact on my business and on my life. I can see opportunities where previously I didn't and feel that I can tackle any challenges and succeed.
- I felt encouraged, enthusiastic and positive, and I also addressed some of the issues about my business that I had been avoiding. Many things I didn't realize I was concerned about, became minor problems, easily solved. I developed strategies and practices that helped me in time management and business practices.
- Change in attitude, more confidence, more relaxed, better time management, felt less isolated, greater faith in my ability to analyze and solve problems for myself.
- I had entrenched patterns of thinking and didn't realize that I felt overwhelmed by the administrative and time-management issues. When solutions were proposed by my mentor, I felt renewed confidence, enthusiasm and energy.

Ineffective

- I had a specific set of issues I wanted to be addressed and the mentor wanted to follow his . . . plan of action.

TABLE 5.17 Unanticipated Outcomes

Effective

- I did not expect to affect career choice and direction, but it was a positive outcome.
- Greater confidence and a different view of my working life
- Understanding the limitations in my knowledge and experience

Ineffective

- . . . to be able to work together in future

---

gram is evaluated only in terms of its program goals, the value or otherwise of side effects is implicitly valued at zero and that this is unsatisfactory.

All of the "side effects" or unanticipated outcomes identified by mentees involved in both effective and ineffective partnerships were positive. They were therefore, in Scriven's terms, not fatal to the program or participants.

The most significant unanticipated outcome evident in the quotational data was that identified under the nature and quality of program structure and content dimension as an unintended consequence of the email facilitation messages. As discussed previously, the messages sometimes had the effect of making mentees feel they were not completing the program "properly." While not fatal, this unanticipated difficulty potentially seriously impacts on effectiveness for mentees.

CONCLUSIONS

Nature and Quality of Mentee/Mentor Relationship

The comparative analysis provided evidence in support of a link between the nature and quality of the mentee/mentor relationship and effectiveness. The comparison of data for mentees involved in effective and ineffective partnerships suggested that effective relationships were valued for their strong bond, mutual respect, a mentor's interest in the mentee's agenda, and good communication. The effective mentoring partnerships were also characterised by a diversity, breadth, and depth in advice and support/s provided, and the use of the mentor as an impartial and neutral sounding board. This analysis is consistent with the small business mentoring literature that suggests that the mentee/mentor interaction "form[s] the cornerstone of subsequent activities... and is... most significant" (Devins & Gold, 2000, p. 254).

The data provided support for Single and Single (2005) and Bierema and Merriam's (2002) findings that the use of the mentor as a neutral and impartial sounding board is linked to effectiveness, and that exchanges in effective relationships are egalitarian in nature. The data also substantiated O'Neill (1998) and Kram's (1980 cited in Noe 1988 p. 459) findings that effective mentoring and e-mentoring is characterized by diversity in the types of support and advice provided.

Nature and Quality of Program Content and Structure and Adaptation to Mentees' Needs

The comparative analysis provided evidence that those involved in effective partnerships were involved in matches where mentees perceived some similarity with their mentor. In contrast to mentees participating in ineffective partnerships, those involved in effective partnerships generally found the program content and structure useful. Mentees in effective partnerships adapted the content and structure, and integrated the program into their business activities. There was evidence to suggest a link between effectiveness and adaptation of the program's structure and content to the mentees' needs, and support for a link between effectiveness and goal-setting throughout the program. There were also quotational data to suggest a positive link between integration of the program with business activities and effectiveness. Mentees involved in effective and ineffective partnerships expressed some concerns about the length of, and demands imposed by, fortnightly facilitation messages from the host which potentially impacted on effectiveness.

Harris et al. (1996; cited in Single & Single, 2005, p. 303), O'Neill et al. (1996) and Single and Muller (2001) suggest that the maintenance of mentoring relationships across email benefits from the use of structure or support. The quotational data provided support for these findings in this context.

There is scope for further research into the specific programmatic features that influence the effectiveness of structured e-mentoring. There is also much scope for investigating the influence of contextual variables to refine the link between effectiveness and Information quality found in this qualitative study, and to further explore for whom and why effectiveness is linked to aspects of the program structure and content and its adaptation to suit the needs of mentees.

Use

In line with Bierema and Merriam's study (2002), which found frequent interaction to be critical to the success of a mentoring partnership, the comparative analysis provided evidence in support of the proposition that those involved in effective partnerships engaged in regular contact with their mentor, providing initial evidence of a link between effectiveness and interaction frequency. The quotational data, however, provided a basis for the dimension of use being problematized to consider the quality and content of the email exchanges alongside interaction frequency as critical to effectiveness evaluation.

The advantages and disadvantages of using email as the primary means of communication between mentee and mentor were largely confirmed in this context in the cases observed. Disadvantages included the lack of cues associated with face-to-face communication and negative impact on communication and learning compared to face-to-face mentoring in some instances.

The advantages identified in the literature were confirmed as the egalitarian nature of exchanges, the capacity for email-based communication to remove obstacles such as geographic dispersal and time constraints, and email providing a basis for sophisticated exchanges between participants, thereby improving the chances of higher learning (Kanuka, 2005; Bates, 1995; Garrison & Anderson, 2003; McGreal, 1998). The quotational data provided support for the additional benefits identified in the research such as the value of an impartial sounding board (Single & Single, 2005).

On the basis of the quotational data, it is possible to infer a link between the dimension of use and effectiveness in the context of structured e-mentoring.

User Satisfaction

While user satisfaction alone is an insufficient measure of effectiveness, there is support for the relationship between user satisfaction and effectiveness (Gatian, 1994; cited in Myers et al., 1998). The comparative analysis provided confirmation for such a link in the context of structured e-mentoring.

Impact

The content and program structure were intended to support the diverse forms of learning required by those in small business, and to allow for, in Devins and Gold's (2000) terms, unpredictable learning pathways. Because this is exploratory work, the indicators of "evidence of learning," "meeting mentees' needs," and "general benefits" were broad to accommodate the diverse and unpredictable forms of learning, the range of needs being met (or not), and general benefits (or lack thereof). The approach ensured that interviewees were specifically asked to identify unexpected as well as anticipated benefits, and to identify possible evolving (Patton, 1986) and long-term benefits (Clutterbuck, 2003). This approach yielded rich and complex data, which provided a basis for identifying patterns in responses and proposing an initial linkage between impact and effectiveness.

The preceding comparative analyses provided evidence in support of the proposition that those involved in effective partnerships engaged with the program as a learning framework and reported benefits in the form of positive learning outcomes, wide-ranging business and other benefits, long-term benefits, their needs being met, and anticipated as well as unexpected outcomes achieved. In contrast, those involved in ineffective mentoring partnerships described more limited, if any, positive learning outcomes, and fewer and less compelling benefits or outcomes.

SUMMARY

The quotational data set out in Tables 5.2 to 5.17 provided a basis for confirming links between each of the dimensions and effectiveness in the context of structured e-mentoring for small business. In turn, this enabled inferences to be drawn about the determinants of effective structured e-mentoring, as set out in Figure 5.2. The use of the term "determinant" is qualified to denote influence or linkage rather than direct causality in this context. Linkages between each of the factors used to operationalize the effectiveness dimensions were empirically established. Because of the con-

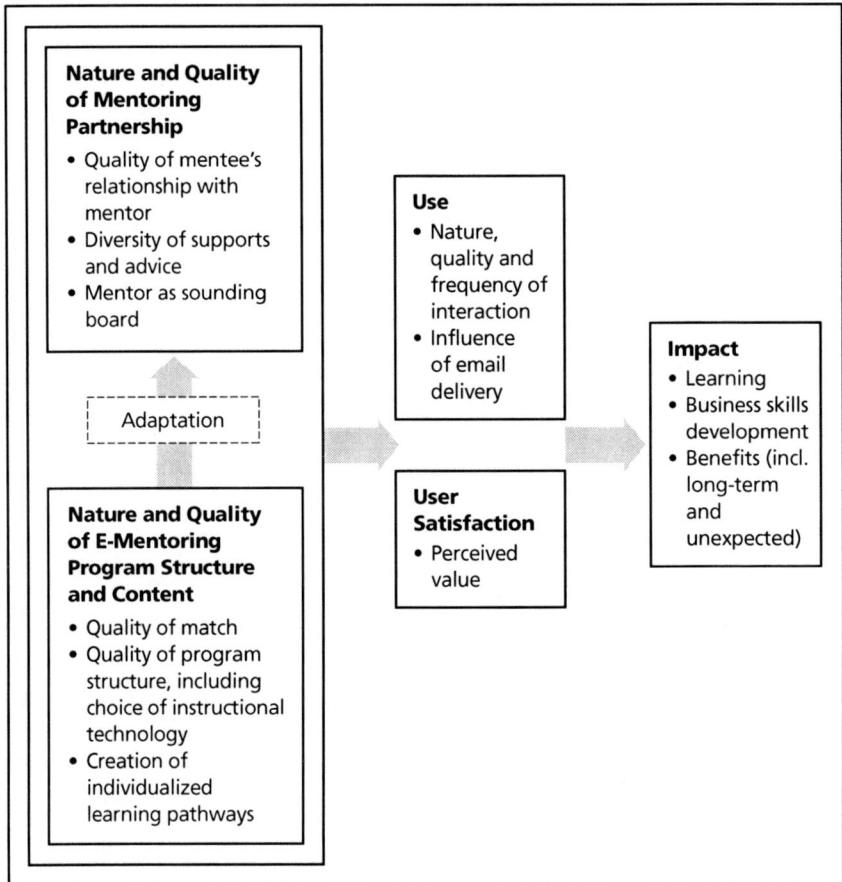

Figure 5.2 Rickard model of structured e-mentoring effectiveness arising out of examination of e-mentoring practice for professionals operating as self-employed contractors (as derived from the DeLone and McLean model of IS success)

tingent nature of these linkages and their specificity to the context in which they occurred, making generalizations, in Clutterbuck's (2003) terms, about "classes of mentoring phenomena" is problematic. However the interpretation and inferences drawn from the data can be considered credible and valid "context-bound extrapolations" as set out by Patton (1990, p. 491).

With linkages established between effectiveness and each of the measures used to operationalize the dimensions set out in the model, positive outcomes in the following areas were established as critical to success: (a) the quality of the mentee's relationship with the mentor, (b) the diversity of support and advice sought and received by the mentee, (c) the use of the mentor as a sounding board, (d) the quality of the match and program structure, (e) the creation of individualised learning pathways, (f) the quality and extent of adaptation of program content, (g) the nature, quality and frequency of interaction between mentee and mentor, (h) email delivery, (i) perceived value, and (j) learning, business skills development, and benefits including long-term and unexpected benefits.

REFERENCES

Bates, A.W. (1995). *Technology, open learning and distance education.* New York: Routledge Studies in Distance Education.

Bierema, L.L., & Merriam, S.B. (2002). E-mentoring: Using computer mediated communication to enhance the mentoring process. *Innovative Higher Education, 26*(3), 211–227.

Boyle-Single, P., & Muller, C.B. (2001). When email and mentoring unite: The implementation of a nationwide electronic mentoring program. In L.K. Stromei (Ed.), *Creating Mentoring and Coaching Programs* (pp. 107–122). Alexandria, VA: American Society for Training and Development.

Clutterbuck, D. (2003). *The problem with research in mentoring.* Retrieved July 11, 2005 from http://www.coachingnetwork.org.uk/ResourceCentre/Articles/View Article.asp?artId=82

DeLone, W.H., & McLean, E.R. (1992). Information systems success: The quest for the dependent variable. *Information Systems Research, 3*(1), 60–95.

DeLone, W.H., & McLean, E.R. (2002). *Information Systems Success Revisited,* paper presented to 35th Annual Hawaii International Conference on System Sciences, Hawaii.

Devins, D., & Gold, J. (2000). "Cracking the tough nuts": Mentoring and coaching the managers of small firms. *Career Development International, MCB University Press, 5/4/5,* pp. 250–255.

Garrison, D.R., & Anderson, T. (2003). *E-Learning in the 21st century: A framework for research and practice.* London: Routledge/Falmer.

Gatian, A.W. (1994). Is user satisfaction a valid measure of system effectiveness? *Information and Management, 26*(3), 119–131.

Hartshorn, C., & Parvin, W. (1999). *Teaching entrepreneurship: Creating and implementing a naturalistic model.* Retrieved ??, http://www.graduatesforgrowth.org.uk/images/naturalisticmodel.doc

Kanuka, H. (2005). An exploration into facilitating higher levels of learning in a text-based internet learning environment using diverse instructional strategies. *Journal of Computer-Mediated Communication, 10*(3), 8.

Kram, K.E. (1980). 'Mentoring Processes at Work: Developmental Relationships in Managerial Careers', Doctoral dissertation. Yale University.

McGreal, R. (1998). Integrated distributed learning environments (IDLEs) on the Internet: A survey. *Educational Technology Review, Spring/Summer,* pp. 25–31.

Myers, B.L., Kappelman, L.A., & Prybutok, V.R. (1998). A comprehensive model for assessing the quality and information systems function: Toward a theory for information systems assessment. In E.J. Garrity & G. Lawrence Sanders (Eds.),

Information Systems Success Measurement (pp. 94–121). Harrisburg, PA: Information Technology Management, Idea Group Publishing.

Noe, R.A. (1988). An investigation of the determinants of successful assigned mentoring relationships. *Personnel Psychology, 41*, 457–479.

O'Neill, D.K. (1998). Engaging science practice through science practitioners: Design experiments in K–12 telementoring. Doctoral dissertation. Northwestern University, Illinois.

O'Neill, D.K., Wagner, R., & Gomez, L.M. (1996). Online mentors: Experimenting in science class. *Educational Leadership, 54*(3), 39–42 .

Patton, M. (1986). *Utilisation-focused evaluation* (second edition). Thousand Oaks, CA: Sage.

Patton, M.Q. (1990). *Qualitative evaluation and research methods.* Thousand Oaks, CA: Sage.

Perren, L. (2002). *E-mentoring of entrepreneurs and SME managers: A review of academic literature.* Small Business Research Unit, University of Brighton.

Rickard, K.M. (2007). *E-mentoring and information systems effectiveness models: A useful nexus for evaluation in the small business context?* October 2007, Doctoral dissertation. Victoria University of Technology.

Scriven, M. (1993). *Hard-won lessons in program evaluation* (vol. 58). New Directions for Program Evaluation. Memphis: Jossey·Bass.

Single, P., & Muller, C.B. (2001). When email and mentoring unite: The implementation of a nationwide electronic mentoring program. In L.K. Stromei (Ed.), *Creating mentoring and coaching programs* (pp. 107–122). American Society for Training and Development.

Single, P., & Single, R.M. (2005). E-mentoring for social equity:Review of research to inform program development. *Mentoring and Tutoring, 13*(2), 201–320.

CHAPTER 6

EXPANDING THE BUSINESS VALUE OF MENTORING WITH A WEB-BASED PROCESS

Randy Emelo and Tom McGee

ABSTRACT

Mentoring can impact employee development at every critical transition point in the employee lifecycle. Web-based mentoring uses technology to amplify this impact for individuals and organizations, broadening the reach of programs and the concepts of what mentoring can be. The practice positively affects employee engagement, retention and productivity, as seen in research studies by Triple Creek Associates. Building momentum for web-based mentoring and expanding its use are critical to creating an effective program. This paper includes practical advice and best practices, as well as case studies of how other organizations have achieved such success.

THE MANY USES OF MENTORING

The practice of mentoring today goes well beyond relationships where one senior leader mentors the person who will one day take over in their job. Today's mentoring practices allow for short-term relationships focused on one press-

Virtual Coach, Virtual Mentor, pages 109–127

ing issue, as well as longer term relationships based on job responsibilities. As people begin to rethink how they can apply mentoring in their lives, they also start to see the many ways in which they can conduct their relationships.

Mentoring can be used to enhance or accelerate employee development at every critical transition point in the employee lifecycle, from onboarding to senior level succession planning. It is can also be used to increase knowledge retention, job integration, and productivity when used with almost any leadership or talent development process. When you take the traditional practices of mentoring and blend them with technology such as the Internet and online social networking avenues, you amplify the impact mentoring can have on an organization. Web-based mentoring leverages the positive outcomes related to retention, job integration, and productivity by expanding the number of people who can participate in mentoring, decreasing administrative costs for mentoring programs, and increasing the organizational benefits of individual mentoring programs. These all positively impact the business value of mentoring, making it an even stronger factor in the success of an organization in the marketplace.

While many initiatives can be administered in a traditional train and match fashion, using a technology-assisted process can change the way in which organizations conceive programs to broaden or enhance their organizational impact. Take, for example, high-potential programs. In traditional programs, senior leaders are often expected to carry the entire mentoring load for up to three mentees, on top of their already pressing workloads. This can stymie not only the enthusiasm senior leaders have for the program, but can also cause them to not participate. Secondly, the complex learning needs of high-potentials are seldom satisfied by a single person, no matter what the mentor's talents, skills, and experiences may be. Expecting one person to do it all and be it all only helps to set them up for failure. Lastly, some traditional programs assign mentors and mentees to one another, which allows the perceptions of administrators to form boundaries around the relationship. This can limit the learning potential of the pair from the outset.

In comparison, a web-based process enables the recruitment of a broad cadre of mentors from various parts of the organization who bring with them a wide range of expertise. This frees senior leaders from carrying the entire load for the development of high-potentials. It also enables mentees to have multiple mentors, either in sequence or concurrently, allowing them to address their most pressing learning needs at their own pace. Most importantly, a web-based process can allow mentees to choose their own mentors from a large, diverse pool of candidates, which fosters a free range of exploration and potentially less politicized relationships. All of this puts more responsibility for the learning in the hands of the high-potentials, testing their initiative and commitment. To help perpetuate learning, high-potential program graduates should serve as mentors to the next group of learners, which helps to embed mentoring further into the leadership culture.

This web-based approach is in practice at a leading banking institution that has offices in 22 countries and over 8,000 employees. In the highly competitive world of investment banking, companies invest large sums of money in new graduates—recruiting them, training them, and helping them build skill in the banking environment so they can become future leaders of the organization. This company needed an onboarding process that targeted early-stage high-potentials. Retaining these new hires for the long term (generally over four years) posed a problem; attrition rates in the investment banking industry ranged from 15 to 17%, according to the organization's mentoring program champion. The bank believed mentoring could help them attain their goals to reduce attrition among graduates, improve employee morale, speed up graduate training, help graduates align with the corporate culture, and increase graduate productivity. Instead of employing traditional methods, they approached a broad range of potential mentors with a wide range of expertise. As a result, they had almost 380 mentors sign up during their initial recruitment phase, well more than they needed for their 250 graduate mentees. This situation enabled mentees to pick from a broad choice of mentors and led to a high degree of satisfaction and learning among program participants. Additionally, having a web-based process enabled them to enroll new hires into the program on a quarterly basis, instead of waiting for the annual cycle associated with most high-potential programs.

Technology opens up a whole range of program design options, as well as exposes the mentoring process to virtually all knowledge workers in an organization. Table 6.1 provides suggestions on how mentoring can be used to enhance employee development, the anticipated organizational impact that mentoring brings to these processes, and key program design considerations that may make that particular mentoring process more effective. When done right, mentoring can be used to enhance or accelerate employee development at every critical transition point in the employee lifecycle.

A web-based process fosters a more creative approach to mentoring programs. For example, a Fortune 100 healthcare services company initially looked to launch a diversity-focused mentoring program, with a traditional approach of recruiting women and minorities as mentors and mentees. After reflecting on the new possibilities afforded by a web-based approach, the diversity committee decided to be the sponsoring organization for a much broader initiative that targeted all employees as potential mentors and mentees. Instead of being seen as an exclusive network for women and minorities, the mentoring program became an inclusive initiative that benefits the enterprise.

TABLE 6.1 Mentoring Integration Points

Core Organizational Development Process	Mentoring Impact on Organization	Key Mentoring Program Design Considerations
Corporate Restructuring	Increased organizational loyalty and change tolerance during critical transitions.	• Recruit mentors broadly from early adopters to help with enculturation of new processes and structures. • Recruit mentees from among those who need help with integrating and adjusting to changes.
Development and Performance	Rapid integration of learning into job actions.	• Recruit mentors broadly. • Ask supervisors to suggest to potential mentees that they find a mentor to help with performance gaps and developmental needs.
Diversity and Inclusion	Increased employee commitment and formation of supportive developmental relationships.	• Recruit mentors broadly. • Allow mentees to choose mentors based on learning needs, career aspirations and cultural considerations.
E-learning or Training Integration	Rapid integration of learning into job actions.	• Recruit course graduates as mentors. • Have mentees select a graduate mentor post-course to help integrate learning into their jobs.
Employee Onboarding	Rapid integration of learning into job actions and assimilation into organizational culture.	• Require each mentee to have two mentors: one a recent hire in the same department and a level or two above the mentee to help with basic orientation issues; the other a more seasoned leader who can impart organizational vision and culture in a short-term relationship.
Employee Retirement	Transfer of critical tacit knowledge and increased commitment to organization.	• Recruit mentors from among retired and soon-to-retire employees. • Recruit mentees from populations where it is critical to transfer outgoing tacit knowledge.
General Leadership Development	Continuous learning of critical leadership skills and processes.	• Recruit mentors with a wide range of expertise and experience. • Have mentees use the system to find mentors as needed for personal and professional development.

Category	Benefit	Recommendations
High-Potentials	Well-rounded and well-networked leadership pipeline.	• Recruit mentors from among current senior leaders who have broad areas of expertise. • Require mentees to have more than one mentor and to be mentors themselves as part of their development.
Knowledge Management Processes	Transfer of critical tacit knowledge.	• Recruit mentors broadly. • Ask mentees to find mentors across silos, in areas of critical need, or for personal or professional growth and development.
Leadership Universities	Rapid integration of leadership learning into current leadership responsibilities.	• Recruit mentors from among current leaders who have broad areas of expertise. • Suggest mentees have more than one mentor and be mentors themselves following course completion.
Merger or Acquisition Integration	Increased organizational unity and cohesion, as well as formation of productive knowledge alliances.	• Recruit mentors and mentees broadly from both organizations to facilitate the formation of new relational and professional networks.
New Job Orientation	Just-in-time learning of critical job requirements.	• Require mentees to find a mentor with experience in a similar position.
New Leaders or Managers	Just-in-time learning of critical job requirements.	• Recruit mentors broadly. • Require mentees to find a mentor with experience in a similar position.
Personal and Professional Development	Continuous learning of critical job skills and exploration of possible career paths.	• Recruit mentors with a wide range of expertise and experience. • Have mentees use the system to find mentors as needed for personal and professional development.
Relational Networking	Broad relational connections and personal development.	• Recruit mentors broadly. • Have mentees find mentors across silos or in areas of interest for personal or professional growth and development.
Succession Planning	Well-rounded and well-networked senior leaders.	• Recruit mentors from among current senior leaders who have broad areas of expertise. • Suggest mentees have more than one mentor over the life of the program.

For those organizations that prefer third-party matching, technology can also play a unique role in mentoring initiatives of this type. Technology can be used to simplify and assist in the matching process, deliver just-in-time training and support to participants, and capture metrics throughout the life of the program. Regardless of how organizations choose to leverage it, technology can make a meaningful impact on any and all mentoring initiatives.

WEB-BASED MENTORING AND EMPLOYEE ENGAGEMENT

In *Follow this Path: How the World's Greatest Organizations Drive Growth by Unleashing Human Potential* (Coffman & Gonzalez-Molina, 2002), authors Curt Coffman and Gabriel Gonzalez-Molina analyze a Gallup Organization study of over 10 million customers, 3 million employees and 200,000 managers. Based on their research, they established a direct link between employee engagement and core business concerns such as turnover, productivity, customer metrics and profitability.

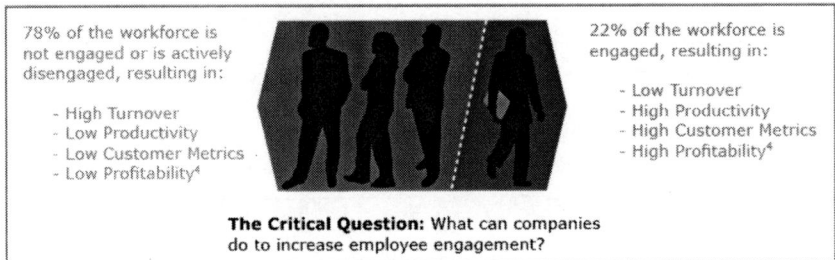

78% of the workforce is not engaged or is actively disengaged, resulting in:

- High Turnover
- Low Productivity
- Low Customer Metrics
- Low Profitability[4]

22% of the workforce is engaged, resulting in:

- Low Turnover
- High Productivity
- High Customer Metrics
- High Profitability[4]

The Critical Question: What can companies do to increase employee engagement?

©TCA

To date, very little research has focused exclusively on web-based mentoring to assess its impact on employee engagement and to determine whether it delivers similar results to traditional mentoring approaches (both formal and informal). To help fill this void, Triple Creek Associates surveyed their clients who use Triple Creek's web-based mentoring process to evaluate its impact on employee engagement, retention, and productivity.

In the spring of 2007, Triple Creek sent a survey to 738 participants at five client organizations. This 10-question survey used a 6-point Likert scale (Strongly Agree to Strongly Disagree) and was designed to gauge the connection between a web-based mentoring program (Open Mentoring®) and eight factors identified as key in terms of employee engagement (see Table 6.2). This attitudinal survey was designed to measure the perceptions of program participants because their perception is the basis for their degree of engagement.

TABLE 6.2   Eight Employee Engagement Factors

- Having adequate resources
- Having development opportunities
- Making personal connections
- Receiving constructive feedback
- Having clear goals
- Making a personal contribution
- Understanding task significance
- Feeling pride in one's company

Of the 738 surveys electronically distributed, 125 were completed and returned. Overall, results showed that a web-based mentoring program can positively impact employee attitudes toward all eight themes of employee engagement. For example, 89% of respondents said mentoring allowed them to contribute to the success of their organization, while 83% felt mentoring helped them enhance skills needed to perform their jobs. Highlighting web-based mentoring's impact on employee engagement, 87% said they felt more committed to their organization because of mentoring.

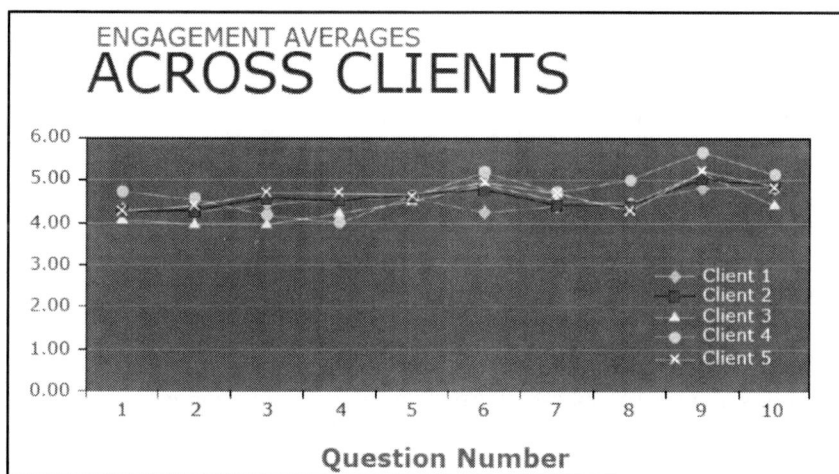

ENGAGEMENT AVERAGES
ACROSS CLIENTS

©TCA

The similar results across all five organizations suggest that organizational differences do not impact the effectiveness of web-based mentoring as much as often believed. In addition to the general results suggesting that web-based mentoring positively impacts attitudes critical to employee engagement, other findings reinforce the effectiveness of web-based mentoring. For example, distance mentoring relationships showed statistically identical results to face-

TABLE 6.3   Employee Engagement Research Summary

Question	Issue Addressed	Results[a]	Percentage Positive
Q1 My experience in the mentoring program has helped me clarify my role at work.	Role clarity	4.37	86%
Q2 The mentoring program helped me enhance skills I need to perform my job.	Resources	4.32	83%
Q3 The mentoring program gave me a feeling of control over my development needs.	Autonomy	4.38	82%
Q4 Due to my experience in the mentoring program, I have felt pride in my work.	Pride in work	4.35	87%
Q5 Mentoring has allowed me to contribute to the success of the company/organization.	Task significance	4.62	89%
Q6 Because my company/organization offers a mentoring program, I feel they value my development.	Development opportunity	4.79	89%
Q7 Because of my mentoring experience, I feel more committed to my company/organization.	Personal connection	4.65	87%
Q8 Due to the mentoring program, I have received valuable feedback.	Feedback	4.61	78%
Q9 The mentoring program helped me develop a positive relationship with another individual in the company.	Personal connection	5.23	90%
Q10 The mentoring program has been a valuable learning opportunity for me.	Development opportunity	4.86	89%

[a] 6-point scale

to-face relationships in their impact on employee engagement. Additionally, mentors and mentees had similar responses, indicating that web-based mentoring positively impacts the engagement of all participants (Table 6.3).

WEB-BASED MENTORING'S IMPACT ON RETENTION AND PRODUCTIVITY

Given mentoring's long association with positively impacting retention and productivity, Triple Creek conducted research in the fall of 2007 to evaluate how web-based mentoring influenced these two key business values. Incorporating client input, Triple Creek designed a survey to measure the connection between Open Mentoring® programs and key indicators of employee retention and productivity; seven factors were selected. Additional questions rated satisfaction with the mentoring experience and the amount of total time invested on a monthly basis. Table 6.4 lists the questions, the issue they addressed, and results.

This survey captured respondents' perceptions and attitudes, which are important to measure because they provide the basis for users' actions and the resulting organizational impact. Distributed electronically to 879 participants across 11 organizations, Triple Creek received 361 responses, a response rate of 41%.

Eighty-five percent of respondents saw a positive connection between the mentoring program and factors that influence retention. Similarly,

TABLE 6.4 Retention and Productivity Research Summary

Question	Issue Addressed	Results	Percentage Positive
Q1 Please rate your satisfaction with your mentoring experience.	Satisfaction	3.93[a]	76%
Q2 The mentoring program demonstrates my organization's commitment to provide career options and opportunities.	Retention	5.02[b]	94%
Q3 My mentoring experience has positively influenced my desire to stay at the organization.	Retention	4.41[b]	83%
Q4 Because of my mentoring experience, I speak positively about this organization to others.	Retention	4.50[b]	83%
Q5 Because of my mentoring experience, I see myself continuing in this organization for the foreseeable future.	Retention	4.35[b]	79%
Q6 Because of the mentoring program, I have become more effective at my job.	Productivity	4.26[b]	80%
Q7 The mentoring program helped me better understand how to access resources in the organization.	Productivity	4.29[b]	78%
Q8 My mentoring relationship provides opportunities to develop new job skills.	Productivity	4.48[b]	82%
Q9 How much time do you spend mentoring each month?	Time spent	2.12	About 1 hour a month

[a] 5-point scale
[b] 6-point scale

80% of respondents saw a positive connection between the mentoring program and facets of productivity. The most telling result, however, was that 93% of respondents indicated the mentoring program demonstrated their organization's commitment to providing career options and opportunities. This incredibly high result points to the contribution mentoring programs can make in developing a culture of learning and opportunity. When used in a web-based format, mentoring can be delivered across the enterprise, leveraging these benefits to hundreds or even thousands of participants instead of the few dozen normally involved in formal programs. The business value of mentoring can expand to the point where mentoring becomes a core delivery model for the talent management and personnel development processes of companies, organizations, government entities, and the like.

Interestingly, participants in this survey who invested at least one hour per month in mentoring-related activities showed significantly higher ratings overall. Those who spent less than 30 minutes per month on their mentoring relationships, nearly one-third of respondents, rated all questions on the 6-point scale lower by almost one full point on average. Based on these results, organizations seeking to impact retention and productivity should encourage mentoring participants to invest one to two hours per month in preparation and meeting time.

This encouragement to invest time and energy into the relationship is particularly important in programs that involve distance relationships. Sixty percent of the distance mentoring participants indicated that they spent 30 minutes or less per month on the relationship, a disproportionate percentage when compared to people in face-to-face and blended relationships who invested the same amount of time (16% and 20% respectively). For programs that support distance relationships, both parties should be encouraged to commit one to two hours per month on their relationships to maximize the impact of the mentoring experience; of note, those in distance relationships who spent one to two hours per month on their relationships had statistically identical results to the face-to-face group, highlighting the positive impact a little more time can make. By supporting participants with the tools, knowledge, and best practices associated with distance mentoring, champions can open up mentoring opportunities to a significantly larger audience, all with the confidence that their efforts will make a difference.

Both studies examined in this chapter show that using web-based processes to promote and manage mentoring programs can yield positive results on employee engagement, can positively impact key attitudes on retention, and can increase productivity by providing key development opportunities. Organizations can take advantage of these realities by embedding mentoring into virtually any personal and professional development process where knowledge is best transferred in a relational context. The possibilities for knowledge sharing, particularly in large or geographically distributed organizations, become endless.

BUILDING MOMENTUM FOR WEB-BASED MENTORING

People generally enjoy learning from each other and satisfaction rates within most mentoring programs are typically very high. Unfortunately, many mentoring initiatives either die or remain in isolated pockets, with little influence on the learning needs of the organization. Why do these initial mentoring success stories fade and lose momentum? What must organizations do to leverage the power of a web-assisted process across the enterprise?

Momentum Killers

While initial success and energy may be high, maintaining that momentum can be a challenge. Impeded progress often comes in the form of these momentum killers.

- *Start small, stay small, die small.* Too often, what started as a successful pilot program gets stuck in that form and lives out its life with minimal impact on the enterprise. This is not to say that individuals did not benefit greatly or that some corporate goals were not achieved, but the potential for the program is never fully realized. The unfortunate reality is that most pilot programs make it almost impossible to expand mentoring because they are built around non-scalable processes such as intensive classroom training or extensive personality profiling combined with hours of expensive third party matching. When the small pilot program succeeds, it is assumed that all of these factors were critical to the success of the program. With this mindset, training budgets and the spare time available to run the program become permanent limitations, and what started as a small pilot is now the "mentoring program." Almost by definition, an organization with thousands of employees will never see a program or process that involves only 10 to 60 people as a part of the core development track.
- *Get it up and running, then set it on auto-pilot.* Some programs start with enough critical mass to be successful, but assume the program will continue without much ongoing effort. While it is true that initial implementation is the most time-consuming part of a mentoring program, a significant amount of effort must be put into reinvigorating and expanding the program. It is also too common that very

successful programs, with hundreds of participants, lapse into disuse over time for a number of reasons:

– Lack of regular intensive recruiting of mentors and mentees.
– A change in mentoring champion without adequate transfer of vision and skills.
– Loss of senior support because of a job change at that level.
– Lack of publicity to keep the profile of the successful mentoring program high in senior leaders' eyes.
– Resistance to change or improve the mentoring experience for participants, even when based on their feedback.

• *Focus solely on making your own program successful.* There is a saying that "a rising tide lifts all boats." Expanding the use and profile of mentoring across an enterprise will make all mentoring programs more successful and less likely to be cut. However, many mentoring champions see helping other leaders discover the value of mentoring as a burden and distraction to their job. They also figure that senior leaders judge the viability of their program solely on the anecdotal comments of participants. The pressure to succeed in one's direct responsibilities can leave little time to look at mentoring from a larger perspective. However, failure to gain senior support and expand the use of mentoring across the enterprise will keep it marginalized in the talent management arena.

Building Mentoring Momentum

To create a successful mentoring program, you need to build momentum and keep it going as the program grows. Success perpetuates success, and the following suggestions can help you build and maintain momentum.

1. *Look for ways to promote and publicize mentoring on a quarterly basis, even if you have a time-bound program.* Both ongoing, self-directed mentoring programs and time-bound formal mentoring programs can lose momentum through failure to communicate enough to keep mentoring in front of key stakeholders—participants, senior leaders, peer leaders, and the general population of the organization. Interestingly, more structured programs tend to ignore the need to communicate with peers who could champion mentoring in their own sphere of influence. Programs of all types could benefit from an annual or semi-annual recruiting/recognition cycle where the profile of mentoring is raised across the enterprise. Many short-term strategies can also be used to raise the awareness of mentoring throughout the enterprise.

- Present brief commercials at other training events where people can see the benefits of having a mentor with whom they can discuss what they are learning. This may not only cause some to sign up for a mentoring program, but it may also change the way people see mentoring, putting it in a more practical, developmental light.
- Conduct e-briefings on mentoring that not only promote current programs, but also educate people on getting the most of their *informal* mentoring relationships.
- Sponsor roadshows or lunch & learns where mentoring pairs share their experiences and highlight the collaborative and mutually beneficial nature of the relationship.
- Organize town hall meetings where a brief presentation could be followed by a question and answer session on how mentoring has impacted people.
- Use various media (such as podcasts, newsletters and presentations) with affinity groups to show how the vision for mentoring can extend beyond current programs to new thinking about mentoring possibilities.

No matter what short-term strategies are used, the best strategy is to build a broad network of mentoring champions and advocates across the enterprise, embedding mentoring into the leadership culture.

2. *Set quarterly goals that include expansion of the view and use of mentoring beyond your personal program and contact with senior leaders.* A quarterly plan could include some of the following elements to address both implementation and promotional goals.

- Q1
  - Initial mentor and mentee recruitment communication.
  - Optional training events and associated publicity.
  - General announcement about commencement of the program in available media.
- Q2
  - Executive report to senior leaders on initial success of the program (quantitative measures).
  - Optional webinar or lunch & learn on sustaining a successful mentoring relationship, open to anyone who wants learn more about mentoring (participants and non-participants alike).
  - Mentoring newsletter (such as Triple Creek's *Masterful Mentoring*) offered to participants and others in the organization as a way to promote informal mentoring.
- Q3
  - Mentoring survey to program participants to gather metrics and nominate excellent mentors or mentees for recognition by the organization.

- Written and verbal communications with possible mentoring champions in other departments or leaders of other core development processes about the value of mentoring, with the focus on expanding support across the enterprise.
  - Q4
    - Mentoring appreciation luncheon with awards for outstanding mentors and mentees.
    - Executive report to senior leaders on the quantitative and qualitative results of the mentoring survey.
    - Overall report to the organization on the successes of the mentoring program through available media.
    - Initial publicity about upcoming mentoring opportunities.

In reality, most of these goals are a normal part of many programs and would not require an enormous amount of time to complete. Just making the plan is half the battle to make sure your success is documented and communicated to the right people.

Gaining Senior Level Support

Almost all major personal and professional development initiatives must have strong senior executive support to succeed and remain central to talent management strategy. Senior leaders set budgets and control spending on all development programs. They make sure programs align with both the long-term goals of the enterprise and the short-term strategies needed to accomplish the most current objectives. In short, their support is critical to expanding the role of mentoring in the enterprise.

While it seems like an obvious oversight, most mentoring champions have detailed plans for implementing their programs but almost no plan for gaining senior support. It is no wonder many mentoring programs die for lack of budget or support at the senior level. To move leaders from skepticism to vocal endorsement requires cultivation, not unlike farming. Seeds of information must be planted, followed by nourishment through personal contact where vision can be presented and objections addressed. Each organization is different and has a distinct leadership culture that needs to be understood before a plan can be developed. However, the following factors are components that are often needed to gain and sustain support at the executive level.

- *Assess current leaders to find your ideal audience.* It is important to know your audience and form an accurate impression of them. Identify their current or previous experience with mentoring, coaching, or other person-to-person learning processes. You want to locate leaders who believe that developing people to reach their potential increases organizational value. This audience is likely to be more receptive to the view of mentoring you are proposing.

- *Communicate respectfully with leaders to establish contact and gain interest.* Once leaders have been identified, you need to engage in a conversation with them to pique their interest in a mentoring program and what it can mean for the organization at large. There are a number of strategies you can use to gain their attention and, ultimately, their support—from emails and phone calls to individual and group meetings. Table 6.5 provides suggestions on what to include in your communications.
- *Make leadership a "visible" part of the process by having them participate in kick-off events.* Participants appreciate knowing that senior leaders both endorse and are participating in mentoring relationships. Brief testimonials or comments from leaders add credibility to the program and give permission for people to invest time in mentoring.
- *Go to extra lengths to ensure that senior leaders have a good experience.* Senior leaders will only support something that appears well planned and well executed. Taking the time to ensure they have a positive experience as a mentor or mentee will build your credibility and spotlight the program as a worthwhile venture.

TABLE 6.5 Suggested Communications

- Inform them of the latest research indicating the impact of mentoring on the area most valued by the organization.
- Present a short vision statement that is broad, compelling, and integrated into the core business practices of the enterprise.
- Ask for their feedback and input on your preliminary thinking, as well as for further discussion to explore the business impact of mentoring in more depth.
- Engage them around their experiences and views of mentoring; you may find that you have an ally already or you may find out their assumptions about mentoring.
- Provide preliminary ROI information or other business drivers that you are aware of that demonstrate how mentoring is a "win" for them and the organization.
- Get their input on your business case: where it is strong, where they think it needs work, what business drivers you may be missing.
- Do not try to convince them or sell them; if your business case has a strong base, they will see the value and help you enhance that value through their suggestions.
- Acknowledge both positive and negative mentoring experiences they may have had and share with them your experiences.
- Help them recall their own personal examples of positive developmental relationships that may or may not have been identified as mentoring partnerships; they need to see mentoring as a way of learning and growing, not just as a formal program attached to some initiative.

- *Keep them informed.* As visible supporters of the program, senior leaders need to be kept informed of the program's progress. Celebrating successes together helps everyone recommit themselves to the program. Give them quarterly updates with research, data, and ROI information that shows how their efforts and support are paying off. Also, update them on your progress toward reaching program goals to build a sense of unity among the supporters.
- *Link mentoring to engagement, recruitment, retention, productivity, and diversity development, and keep building the business case around achieving those drivers.* This is where many champions make the mistake of assuming that since their mentoring program has survived a budget cycle or two, it is safe and will continue forever. Mentoring champions must constantly gather data from both internal and external sources to validate the growing use of mentoring. They should also continually identify new senior leaders who can carry the torch when other leaders retire or go on to other assignments.

EXPANSION IN ACTION

A leading communication and aviation electronics solutions company viewed mentoring and coaching as critical levers in their overall Leadership Development Roadmap, which was designed to reach the organizational goal of "developing more and more capable leaders" to serve a growing organization of 19,000 employees in 27 countries. The program was targeted originally for developing leadership benchstrength in a succession management program. However, as the organization began to understand better the benefits of mentoring relationships and as rapid knowledge transfer and learning on a global basis became greater business imperatives, a need arose to scale the mentoring program to accommodate more participants in the program—without increasing administration costs but while improving overall program effectiveness. The target audience rose from 131 people to 19,000 (the entire enterprise). With this expansion came issues regarding efficient and effective facilitation of mentoring relationships.

The company used multiple paper and pencil mentoring programs that were successful, but these separate programs became increasingly difficult to manage as the number of participants increased. For 500 participants, 12 administrators were used (one for each business unit represented), and program management required over 50% of some administrators' time. The organization needed a way to consolidate the multiple mentoring initiatives, reduce the number of hours required to administer the programs, and make the mentoring program available to all employees globally. They chose to accomplish these goals through web-based mentoring. A global pilot began in January 2007 that involved 2,256 total participants (out of approximately 2,500 invited), representing the enterprise population from a demographic and organizational perspective. The pilot concluded in September 2007, at which time participants completed a survey that measured the impact of mentoring on such areas as cost savings, productivity improvements, and acting upon an accelerated learning curve in their current role.

Web-based mentoring allowed mentees to solve real-life work problems with their mentors, the results of which manifested in cost savings, increased productivity, and even improved relationships with leadership and/or the participant's team (see Table 6.6).

Results also showed that web-based mentoring helped participants contribute to the success of the company. This occurred through such actions as transferring valuable knowledge, skill, and experience; building or expanding personal networks; and accelerating their learning curve (see Table 6.7). Yet the greatest way mentoring helped participants contribute to the success of the company was through offering a relationship that provides encouragement, motivation, and support to the mentees. This was the highest rated factor for both mentees and mentors (21.8% and 22.4%, respectively). This directly impacts the morale of employees, giving them tangible proof that the company cares about them. That positive impact on morale can lead to increased employee engagement, which can have a very positive and powerful long-term impact on the company.

Additionally, 74% said they would recommend the mentoring program to someone else. With recommendations such as that, the organization can

TABLE 6.6 My Mentor(s) Helped Me Solve a Problem That...

	Number of Responses	Percentage of Respondents
Resulted in cost savings	8	2.2%
Increased my productivity	43	11.9%
Increased the productivity of a function/team	32	8.9%
Improved my relationships with my leadership/team	82	22.8%
Enhanced my external/customer relationships	32	8.9%
Allowed me to evolve my career development plan	121	33.6%
My mentor did not help me solve a problem	17	4.7%
Other (Please comment below.)	25	6.9%
Total Survey Responses	360	

Note: Check all that apply.

TABLE 6.7

	Number of Responses	Percentage of Respondents
Mentee Question: My mentor(s) helped me to contribute to the success of the company by ... (Check all that apply.)		
Transferring valuable knowledge/skill/experience to me	101	17.6%
Building or expanding my network	83	14.5%
Accelerating my learning curve	38	6.6%
Providing me with encouragement/motivation/ support	125	21.8%
Assisting in my professional development	98	17.1%
Helping me understand a different point of view	108	18.8%
I do not feel mentoring has allowed me to contribute to the success of the company	11	1.9%
Other (Please comment below.)	10	1.7%
Total Survey Responses	574	
Mentor Question: Mentoring has allowed me to contribute to the success of the company by ... (Check all that apply.)		
Transferring valuable knowledge/skill/experience to another person	70	21.8%
Building or expanding my mentee's network	66	20.6%
Accelerating the learning curve for my mentee(s)	49	15.3%
Providing my mentee with encouragement/ motivation/support	72	22.4%
Assisting in the professional development of my mentee	64	19.9%
Helping me understand a different point of view	38	11.8%
I do not feel mentoring has allowed me to contribute to the success of the company	5	1.6%
Total Survey Responses	321	

anticipate strong growth as the program moves from pilot to full-scale implementation across the enterprise. Keys to their success include:

- Positioning mentoring as a core leadership development process
- Gaining key senior level support
- Running brief, successful launch events that kept the profile of mentoring high and provided some training and encouragement for participants

- Starting with enough critical mass to encourage and sustain a web-based process
- Gathering useful metrics to validate success and make needed adjustments.

This case study offers just one of many successful applications of technology to the practice of mentoring. In every case, the core value of mentoring is expanded or enhanced through a web-based process that encourages and supports the development of an inclusive mentoring culture—one that can fully harness the accumulated knowledge and wisdom of large or geographically distributed organizations. In today's global knowledge-based economy, an inclusive web-based mentoring process will likely emerge not just as a strategic advantage, but as a strategic necessity.

REFERENCES

Coffman, C., & Gonzalez-Molina, G. (2002). *Follow this path: How the world's greatest organizations drive growth by unleashing human potential.* New York: Warner Business Books.

Triple Creek Associates. (2007). *Triple Creek's employee engagement research.* Denver: Author. Available online: http://www.3creek.com/resources/research/TCA_Engagement_Research.pdf.

Triple Creek Associates. (2008). *Web-based mentoring's impact on retention and productivity.* Denver: Author. Available online: http://www.3creek.com/resources/research/Retention_and_Productivity.pdf

CHAPTER 7

VIRTUAL COACH AND MENTOR SUPERVISION

Edna Murdoch

ABSTRACT

The coaching profession has eagerly embraced communications technology, and this has increased the capacity for rich exchange and increased the range of learning environments. This chapter will look at how both group and individual coach and mentor supervision can occur virtually and what supervisors need to be aware of when working in this way. It will also highlight significant features of using a combination of telephone, email and recorded material in supervision and what is different when we supervise face-to-face. The chapter illustrates some of the key features of coach and mentor supervision and shows how virtual supervision results in a powerful and effective form of professional development for coaches and mentors. In addition, there is insight into how working virtually affects the body/mind and how the energy of the work is carried across the various media. The supervisees referred to in the chapter are members of virtual coach supervision groups run by CSA. They have been asked to comment on their experience of telephone supervision.

The energy fields that we inhabit in relationships exist in time and space wherever we are. So telephone conversations carry these field conditions . . . we are all connected at levels often outside our everyday conscious awareness.

—Fiona Adamson, Coaching Supervision Academy, 2008

Virtual Coach, Virtual Mentor, pages 129–137
Copyright © 2010 by Information Age Publishing

The coaching profession has embraced communications technology intelligently, and this has greatly increased the capacity for learning and for participative and engaged professional relationships. In this chapter, I will look at how virtual coach and mentor supervision—specifically the telephone, email, and recorded coaching sessions—adds up to a powerful and effective form of supervisory experience. We will look briefly at the "field conditions" that make this possible (what's happening in the space between supervisor and supervisee and around them) and at the main features of coach mentor supervision, in order to show what it's like to supervise and be supervised virtually.

FIELD CONDITIONS

In the world of Quantum Physics, we are all connected, all the time and in every place: "At this level we discover that ALL matter is energy—the desk, the car, you, and me. There is no solid boundary between matter and non matter, and both are made up of the same thing—energy, or quantum particles." (Orriss, 2006, p. 2).

Ervin Lazlo, philosopher and Nobel Prize winner, says that the Field is one vast interconnected field of information. So, when we begin to work with an individual or a group, we enter into and increase the energetic space which contains our creativity, thinking and intentions for the work. This begins to happen from the moment a potential employer or supervisee contacts a supervisor and it continues as we converse and contract for the work. It happens whether we are sitting together in an office or if we are talking via the internet or on the telephone. Given that we are connected electromagnetically, the choice to work virtually requires only a slight change in focus and skills; there is no reduction in connection or in professional effectiveness. Indeed, it has been suggested that using the telephone, for example, may even encourage the use of a higher mental function: "the integrative intuitive perception of a quiet, peaceful mind" (Selby, 2004, p. 87).

COACH AND MENTOR SUPERVISION

Supervision is an opportunity to bring someone home to their own mind, to show them how good they can be.
—Nancy Kline, 1999

I want to highlight the main functions of supervision in order to show how virtual working easily carries its main functions and tasks. Supervision is conversation between two experienced professionals where the central focus is on the coach or mentor's work and on how their being affects that work. Supervision explores and clarifies what goes on in various relationships and conversations, and it enables coaches and mentors to be intelligent about creating effective conversation—conversations with organizations, coaches, sponsors and stakeholders, between coach/mentor and coachee/mentee, and within the coach or mentor. There are several models that can guide these explorations and that are useful in gathering data about what subtly and powerfully influences conversation. For example, CSA's Full Spectrum Model of supervision gives a useful overview of the range and scope of supervisory conversation, while Peter Hawkins' 7-eyed model guides the supervision session in detail.

TASKS OF SUPERVISION

Some of the central tasks of supervision include:

- ensuring that standards and ethics are maintained
- supporting coaches and mentors by increasing personal development
- exploring critical moments in their work
- establishing clear contracts
- clarifying boundaries
- deepening coaching presence
- building the Internal Supervisor
- increasing interventions and tools
- developing systemic awareness
- creating experiments through which the practitioner can learn

Attending to these tasks requires that the supervisor has a strong presence and can connect easily, particularly on the telephone. The supervisor also needs to have the capacity for profound listening. Virtual conversations, listening to supervisees' recordings, and using email support allow all of the above to take place efficiently, especially when we understand that "the energy fields that we inhabit in relationships exist in time and space wherever we are."

The telephone also ensures that coaches and mentors have access to the supervisor who is a true match for them. One supervisee comments:

> I have known my supervisor for several years and she feels like an old friend, and yet I have never met her . . . if I use the telephone for my supervision, I can choose the best supervisor for me, not just the one who lives locally. I can even choose someone who lives abroad if I like. This means that the connection is probably stronger because I can choose the right person for me.

PROCESS OF SUPERVISION

Another way of identifying what supervision does, is to think of it as a process of Reflection, Insight, and Support. This way of understanding coach and mentor supervision underlines the fact that reflection enhances "seeing": the seeing into one's practice, the illumination of subtle processes in professional conversations, and the illumination of blind spots in oneself and in one's thinking. Supervision through reflection is then something that the supervisee, whether coach or mentor, takes away—an enhanced view, a super-vision of their practice. This occurs just as well face-to-face; however, supervisees report that the very concentrated thinking that is a by-product of working virtually is especially helpful in deepening reflection and insightfulness. As one supervisee puts it, "Being an introvert, I like working by telephone. I find that what you miss in body language you gain in focus. Curiously, I think I concentrate more on the phone than I do face to face; there are fewer distractions."

SOME BENEFITS AND CHALLENGES OF BEING INVISIBLE.

Coach and Mentor Supervision understands that while the observable business of coaching is going on—meetings, contracting, outlining coaching programs, coaching sessions—it is people who do the talking and thus, who and how we are in the conversation, affects outcomes. This "who and how we are" piece, while being mostly unobservable from the outside, does have significant impact on effectiveness; supervision, like coaching and mentoring, is a relational practice. The telephone is surprisingly useful in this context—coaches are very willing to explore deeply in the capsule of sound that is created on the phone. Neither are they so bothered if they hesitate, blush or are lost for words.

As one supervisor put it, "Supervision on the telephone enables me to access my feelings more than face-to-face. It's more internal—I go inside more; I access inside myself better. Socializing with others interferes a bit. The telephone allows me to become a bit more introspective."

And so for some coaches and mentors, the personal development explorations that are part of supervision may be more effective when conducted virtually.

VIRTUAL PRESENCE

Many of the major professional bodies highlight Presence as a fundamental skill in coaching, as indeed it is in any professional conversation. Presence is often understood only in a face-to-face context and includes awareness of body language. This, of course, gives access to a wide range of information and a particular kind of immediate connection; for example, if I were to observe only the face of the other in a conversation, that alone would provide me with many clues as the conversation proceeds. When we add in the rest of the potential information, via body, movement, touch and environment, it becomes obvious that virtual relating reduces the available information significantly. This view of presence takes into account only the visual and sensual—it does not expand the understanding of presence in relation to the energetic connection or the field to which I referred earlier in the chapter. I want to emphasize that paradoxically, removing visual and sensual clues can help to create a strong presence and that the supervisor's attention to co-creating a lively connection on the telephone is very important. I say this because, in the absence of many of the things that can enhance contact when face-to-face and create full body/mind presence (visual and sensual data), the supervisor needs deliberately to 'reach across' to the coach or mentor in conversation. In essence the supervisor is concentrating on forming the energetic or virtual connection. Sometimes, at the outset of a session, simply sharing information about the weather outside or the room where one is sitting, makes a link between supervisor and supervisee and connects us across the virtual space. This brings something of each person's environment into awareness which helps to create a bridge and establish contact; it gives literal ground to the virtual space.

> Rather than meeting face to face in some impersonal hotel lobby in London we work from our respective offices where I look out into my garden, the River Pinn meadows and an old oak tree and my supervisor looks out to the sea and where—in my mind—the corporate and personal, the abstract and concrete, and the conscious and unconscious meets and is explored in our virtual space where possibility opens up, undiluted by extraneous noise and interruption.

As supervisor, I find that "listening into" the telephone, often with eyes closed, helps me to become wholly present and arrive more effortlessly at the place of "letting go to let come," which, in Peter Senge's (Senge. Scharmer, Jaworski, & Flowers, 2004) thinking, is where the best learning occurs. Getting there requires that we remove ourselves from habitual streams of thought and that we release the first level of thinking—for example, focusing on what the supervisee did last session, and what he might be requiring from this one, how she is looking today—how *I* am looking today! In Senge's model, suspending our usual assumptions and re-directing awareness to include field consciousness allows for learning about what is emerging and invites intuition that will support the more analytical functions of the mind. This supports supervisor and supervisee to come more fully into presence and encourages "seeing from the deepest source and becoming a vehicle for that source." (Senge 2004). I would suggest that working virtually actually enables this to occur more easily, as the full range of sensing and thinking are enhanced by having fewer field phenomena present, for example: sight, sound, touch, movement. A member of a telephone supervision group puts it this way: "Working on the telephone allows me sense of quietness and presence; I can make meaning from other people's experiences. It's a lovely space."

And in the space between sessions, when a supervisor might be listening to a recording of a coaching session, reflecting on it, and emailing additional material to support learning, there is time to process and to identify what will be most useful to the practitioner. The paradox is that all the live visual and physical data that can guide us in professional conversations can also inhibit us, and the absence of this data can actually promote other means of relating and of deepening reflection.

"Listening into the telephone" is a particular form of listening and requires a denser level of concentration and awareness than when we listen face-to-face. For this reason, it is usual that telephone sessions are briefer than face-to-face ones, as virtual listening is very demanding. My experience is that we listen "into," and not just "on" the telephone as we are energetically reaching for and connecting with the other across literal distances. Some of Nancy Kline's (1999) excellent insights about listening and paying attention are particularly relevant here. She writes that the "quality of your attention determines the quality of other people's thinking" (p. 17) and that "good attention to people makes them more intelligent" (Kline, 1999, p. 37). Supervision is a learning journey and so stimulating thinking, and thinking together, are key components of it. As *listening across distance* is so central to the success of virtual supervision, it is important that supervisors become proficient in attention, presence, and deep listening. These meta-skills are essential for the success of virtual work. In the words of a colleague, "When I'm on the phone I can HEAR much more than I can face to face. I am able to focus wholeheartedly on every word choice, intonation, pause that a client makes—and my ability to hear what is under the words is amplified."

BODY/MIND EXPERIENCE IN VIRTUAL SUPERVISION

There is "something very powerful about the voice entering the mind" of the listener (De Haan, 2008). What is different here from working face-to-face is that the distractions of sensual data—environment, clothing, hair, smell, touch, movement of face and body—are all absent. In my experience, this drives concentration even deeper and requires that the practitioner be not only aware of the details of the voice that they are listening to, but that they be prepared to enquire about how something is said: "Was I right that I heard your voice drop when you said that?" The supervisor will also need to check carefully their assumptions regarding how a client is feeling and confirm their perceptions about all parts of the supervisory conversation, throughout a session. These assumptions in face-to-face work are more easily clarified by just watching closely for somatic signals, whereas when working virtually, picking up the subtle clues of speech patterns, pacing, voice levels, and the quality of particular silences in conversation becomes a core skill. These clues more than make up for not having the sensual clues that are present in face-to-face work.

The effect on the supervisor's body/mind system is more likely to be intensified by working in this 'capsule' of telephone conversation. For example, I am more likely to be aware of how my breathing, posture, heart rate, and facial gestures are changing during a session, if I am not distracted by watching for clues in the client's face and if I do not have to disguise my own body signals—grimaces and other facial gestures—which might otherwise be distracting for the client. As one coach puts it:

> On the phone I can frown, scowl, pace . . . do whatever I need to do to focus on the client's message. Face to face I need to edit myself—I need to ensure that my body language and movements do not disturb the client or give away what's going on for me. The very fact that I am aware of needing to edit my physical response means that I am less present to the client's words.

In fact, I often close my eyes in order to concentrate well and so my body's responses to what is being said or to the current silence in the supervisory conversation, are more obvious and I can use them to increase the range of possible interventions available to me. This means that I may be able to explore more readily aspects of the relationship in supervision that are highlighted by somatic intelligence. For example, I might ask, "I notice that when you said that, we both had an intake of breath; what's that about for you?" Or if I experience a strong feeling in a session that I know does not belong to me, I may be guided to ask the supervisee to explore some aspect of their experience more fully, using both body and mind information. It is important that the supervisor constantly clarifies their perceptions and intuitive hunches since they do not have access to the 'live' clues which would offer information and a quick check.

TELEPHONE GROUP COACHING SUPERVISION— AN EXAMPLE

I prepare to supervise a new group of four executive coaches on the phone. I'm slightly apprehensive. There's nothing unusual in that—Bion suggested many years ago that there should be at least two terrified people in every conversation. He was acknowledging how being open to the unknown in any conversation really affects us. Having only voices to connect with initially on this call makes the "unknown" quite real. These coaches have requested telephone supervision as their preferred way of having supervision, and they know each other quite well; they work in different contexts, but live in the same area and belong to the same coaching chapter.

Our preparation has included visiting our various websites and sharing biographies via email. In addition, each coach has had a conversation with me on the phone, and there has been particular attention paid to setting up and agreeing the contract for this work. The contracting has been done largely via email—many emails. We have finally agreed to have an initial session to check the chemistry for everyone and that the methodology suits our purposes. If this goes well, we will contract for six monthly sessions, one and three quarter hours long, and each person will have time in the group to present issues for supervision. The style and nature of feedback has also been agreed—papers on feedback have been sent to participants, followed by comment and suggestion; this is a key ingredient in group coaching supervision as we have to agree on how we share insights with each other and encourage the potential for generative learning. We are ready to roll.

Eighteen months later, this group and I are working well, enquiring into classic coaching issues and over the months; trust and professional care have developed to such an extent that the calls allow for deep personal and professional exploration, for brave encounter with our glitches, and for humor that gives an edge to the work and keeps us light. There have been explicit requests from time to time for deeper sharing or for information about supervision models—this re-contracting has signaled the development of greater group cohesiveness and increased engagement with learning, as the group has progressed and become more comfortable.

Often as we close a morning call, the comment is made that participants feel well set up for the day—the telephone has been the vehicle for connecting our energies and ideas and for ensuring thorough learning about ourselves and our work.

Supervisors working with a group such as this need to manage the sessions particularly well. There can be potential anxiety for the supervisee if their time on the call is not protected properly, or if there are process or contracting issues that are not explored and resolved. Participants on the call cannot see if the supervisor is getting ready to sort out an issue or allocate time appropriately, and it is not possible to put up one's hand and make a suggestion!

I have supervised many coaches and mentors since 1999. Much of this has been conducted on the telephone, supported by pre-session information that is emailed to me and by email follow-up. Supervisees also send sections of sessions for me to listen to, so we are working with three media at once. And while I will not meet many of the practitioners that I work with, our conversations are nevertheless vigorous, challenging, and profound, and they generate significant learning for the coaches and mentors—and for me. When a supervisee has asked to meet me for the first session, he or she may show some reluctance if I suggest continuing to work virtually. I've not yet come across a supervisee who has regretted changing to working via the telephone—indeed, many are surprised by the effectiveness of this medium and delighted that it fits in so well with a busy professional life. The potency, flexibility, and cost-effectiveness of virtual working is an enormous benefit to coaches, mentors, and supervisors, and I am sure that as the virtual landscape develops, there will be more of it.

REFERENCES

Bion, W.R. (1976–1979). The Tavistock Seminars—from talks at the Tavistock Clinic, London.

De Haan, E. (2008). *Relational coaching: Journey towards mastering one-to-one learning.* Hoboken, NJ: Wiley.

Kline, N. (1999). *Time to think.* West Sussex, UK: Ward Lock

Orriss, M. (2006). *Understanding the energetic principles that enable you to quantum coach.* Unpublished paper. Available from www.coachingsupervisionacademy.com.

Selby, J. (2004). *Quiet your mind.* London: Rider.

Senge, P., Scharmer, C., Jaworski, J., & Flowers, B. (2004). *Presence: Human purpose and the field of the future.* Cambridge: Society for Organizational Learning.

CHAPTER 8

E-COACHING

Consideration of Leadership Coaching in a Virtual Environment

Lisa A. Boyce and Gina Hernez-Broome

ABSTRACT

Leadership coaching is an accepted and widely popular method for developing leaders. In order to meet the growth in interest and expanding needs of clients, coaches and coaching organizations are looking for alternatives to traditional face-to-face coaching, including the integration of technology into their programs. Unfortunately, technology supported practices have outpaced the research supporting guidance on how best to manage and support virtual coaching. In order to meet this need, this chapter presents a framework to highlight the critical issues and alternative considerations for coaching within an e-environment.

After accessing his coach's public scheduler, Julius, a highly motivated mid-level manager of a global distribution company, instant messages Nayah requesting an impromptu telecom to discuss a pressing leadership problem. After the phone conversation, Nayah, an associate leadership coach with over ten years of experience developing mid-level and executive leaders, reviews the digitally stored results of Julius's on-line assessments as

Virtual Coach, Virtual Mentor, pages 139–174
Copyright © 2010 by Information Age Publishing
All rights of reproduction in any form reserved.

well as his goals, milestone charts, and integrated outcome data. Nayah has been work-
ing with Julius for nearly three months and while they have a little in common, they are
effectively compatible with well-matched personalities and similar learning styles. Inter-
preting a trend that was illustrated by this latest crisis, Nayah requests the automated
coaching assistant to send Julius a status update, summary of their recent phone con-
versation, and critical feedback on his progress regarding his action plan. In addition,
Nayah identifies an appropriate exercise from the program's dropdown menu and uses
the accompanying template to tailor and guide Julius on using the tool. Julius, upon re-
ceiving the information, utilizes the automated response options to ask a question about
his homework and permits the system to automatically schedule a follow-up synchronous
computer-mediated communication (CMC) with Nayah. The computerized assistant,
recognizing Julius's question, provides an automated response. Julius finds the response
sufficient and begins the homework. With the prompting of automated reminders, Julius
completes and sends his homework as well as an update on the earlier leadership problem
to Nayah before their scheduled CMC coaching session.

Leadership coaching is an accepted and widely popular method for developing leaders in both private and government organizations. This growth in popularity is a reflection of the benefits experienced by organizations and individuals. Organizations report significant benefits from coaching, including improvements in productivity, increased customer service, decrease in customer complaints, greater retention of leaders, increased profitability, and lower costs (Fisher, 2001; Peterson, 2002). Individuals who have participated in coaching programs report improvements in working relationships with immediate supervisors, peers, direct reports, and clients, as well as with teamwork, job satisfaction, conflict reduction, and organizational commitment (Fisher, 2001; Peterson, 2002). However, these benefits do not come cheaply; one-on-one face-to-face leadership coaching requires a significant investment in time, money, and energy. As a result, coaches and coaching organizations are looking for options and alternatives to traditional face-to-face coaching, including the integration of technology into their programs to meet the demands of their clients.

The experiences shared by Julius and Nayah are being played out in coaching relationships today. Unfortunately, technology-supported practices have outpaced the research supporting guidance on how best to manage and support virtual coaching. In order to meet this need, we present a framework to help the individual coach, those responsible for developing and maintaining organizational coaching programs, and trainers of coaches, as well as their equivalents in the mentoring profession conceptualize the critical issues and consider appropriate actions for coaching within an e-environment. Before we present the framework, we will elaborate on our definition of e-coaching, share with you why we think e-coaching is important to the future, and summarize the current state of e-coaching knowledge.

WHAT IS E-COACHING?

Chapter 1 provided a generic definition in order to encompass both e-coaching and e-mentoring: "A developmental partnership, in which all or most of the learning dialogue takes place using e-mail, either as a sole medium, or supplemented by other media." This definition, focusing on the learning dialogue, serves a valuable purpose in framing the book. We present a more comprehensive and precise definition of e-coaching in order to provide a foundation for understanding the critical coaching factors and their interrelationships. This definition will also serve to establish a reference point to adjust the factors and relationships to meet your individual situations and needs.

Unlike e-mentoring, there is not much variation in how e-coaching or its close synonyms (e.g., on-line coaching, blended coaching, distance coaching, telecoaching, or virtual coaching) are defined. In fact, most definitions (e.g., Dixon, 2008; Frazee, 2008; Hernez-Broome, Boyce, & Whyman, 2007; Marino, 2005; Pulley, 2007) incorporate three key elements: 1) a coach-client relationship, 2) the utilization of technology, and 3) the purpose of facilitating client growth. The definition we present also incorporates these elements.

We capitalize on the Center for Creative Leadership's (CCL) definition:

> a "formal one-one" relationship between a coach and client, in which the client and coach collaborate using technology to assess and understand the client and his or her leadership development needs, to challenge current constraints while exploring new possibilities, and to ensure accountability and support for reaching goals and sustaining development. (adapted from Ting & Hart, 2004, p. 116)

In addition to describing the coaching relationship and results, this definition highlights three core processes of coaching—assessment, challenge, and support—and does not limit coaching engagements to face-to-face interactions but suggests a two-way communication that is enabled through the use of technology, such as computer-mediated communications (CMC).

Examining our definition through Julius and Nayah's vignette, it is possible to identify the parallels to the three key elements highlighted above and in particular, appreciate the unlimited ways technology can be applied in the coaching process. For example, Julius and Nayah not only communicate through instant messaging (IM), the telephone, and computer-mediated communication (CMC), the assessments were performed on-line, information summarized and stored digitally allowing access through multiple media, and web-based resources as well as intelligent systems provided coaching support. While this vignette illustrates potential applications of technology in coaching practices, it also illustrates that we need not and should not limit our definition to focus on only the communication technological aspects of e-coaching. We also need to consider the broader implications, both the positive and negative, of utilizing technology and be forward-thinking, since technology is dynamic and continually evolving.

WHY IS E-COACHING IMPORTANT TO THE FUTURE?

The interest in and use of leadership coaching is expanding dramatically, as evidenced by the increase in leadership coaches, leaders being coached, coaching publications, and membership in professional coaching organizations (Bolch, 2001; Boyce & Hernez-Broome, 2007; Ely, Nelson, Boyce, & Zaccaro, 2008; Johnson, 2004, respectively). With this increased attention has come a change in the nature of coaching with internal coaching development programs gaining prominence, junior and mid-level managers seeking coaching, and the purpose moving from a problem-solving focus towards developing high-potential leaders (Elder & Skinner, 2002; Boyce & Ritter, 2002; Corbett, Corbett, & Colemon, 2008, respectively). Leadership coaching has evolved from executive coaching, exclusive to senior leaders at the top of organizations, to an integral component of leadership development programs that support the expanding needs of leaders at all levels.

The demands of attracting, retaining, and developing talent have also impacted the expansion of leadership coaching with leadership coaching now contributing to talent management and employee engagement. Based on data from the Human Capital Institute's Center for Talent Retention (HR Focus, 2008), among the top five reasons that employees engage and stay with their organizations is that they have the opportunity to develop new skills and capabilities and the employees feel appreciated and valued. Conversely, among the top five reasons that people leave their organizations is that they did not receive any coaching or management support. Thus, the need to make coaching cost-effective and scalable to broader audiences is paramount, given the particular challenges related to talent management.

The nature of organizations and work is also changing. Demographic and societal shifts include an aging, more heterogeneous and disparately educated workforce as well as a generation of tech-savvy employees entering the workforce (Howard, 1995). Leaders are required to perform in more cognitively and socially complex, uncertain, dynamic, political, technological, and global environments (Zaccaro & Klimoski, 2001). Further, organizations are becoming less hierarchical, empowering their employees with decision-making responsibility, while at the same time, restructuring and downsizing have shortened the career cycle (Howard, 1995). A recent survey of nearly 250 senior executives by the Center for Creative Leadership (CCL) reinforced and expanded on these themes by identifying organizational trends, the obstacles they create, and their impact on leaders (Criswell & Martin, 2007). The results identified patterns that focused on complex challenges, (i.e., market dynamics, shortage of talent, globalization) and reliance on innovation, leadership development, and virtual leadership and collaboration to overcome such challenges.

As a result of the popularity and evolution of leadership coaching combined with the changing nature of organizations and work, there is a demand for leaders to lead a diverse workforce, develop cross-cultural competencies, be continuous learners, respond swiftly and effectively to situations, and produce results. E-coaching is well suited to address and support these needs of current and future leaders. As depicted in Julius's situation, e-coaching supports a just-in-time performance requirement while he was geographically separated from his coach. But the vignette provides a limited view of the often-cited and potential applications and benefits of e-coaching.

The benefits of e-coaching can be grouped into three client-centered categories, which we refer to as *at your convenience, at your service,* and *at your fingertips. At your convenience* incorporates the advantages technology provides regarding coaching logistics, such as flexibility of location and time and lower costs. *At your service* addresses the ability of the coach to meet the client's dynamic just-in-time needs, and *at your fingertips* encompasses the availability of tailored resource support, including on-line assessments, exercises, and performance metrics. To further illustrate, we have listed many of the cited benefits (e.g., Rossett & Marino, 2005; Marino, 2005; Olson, 2001) associated with e-coaching within these three categories.

At Your Convenience

- Client and coach can be globally separated, as geographical boundaries are no longer a constraint.
- Clients can employ cultural coaches to support their multinational needs.
- Coaches can reach clients not typically able to receive coaching due to inaccessibility (e.g., military deployment, business travel, expatriates, third-world countries).
- Interactive calendars allow clients and coaches to log on and schedule or reschedule sessions.
- Clients can coordinate synchronous coaching, complete homework assignments, and respond to asynchronous feedback according to their dynamic schedules.
- Coaches' responses can also be viewed, stored, and processed by the client whenever it is convenient and appropriate.
- Asynchronous conversations allow clients to pause and reflect on feedback before responding.

- Asynchronous conversations also support advance thinking time, which allows coaches and clients to prepare and focus during a synchronous coaching session.
- Technology makes coaching more affordable. E-coaching reduces costs of travel, time away from the job, and the expense of face-to-face coaching sessions.
- Technology also supports a greater coach-client ratio, enabling a single coach to work with multiple clients with a greater variety of leadership development needs.

At Your Service

- Coaching can occur just-in-time, addressing client issues as they emerge.
- Coaching can occur during real time, providing guidance and feedback as the client performs.
- Coach can act as sounding-board so the client's decision is validated or challenged before being implemented.
- Coaches can tailor reminders as well as provide support and reinforcement without having to be physically present.
- Coaches can provide immediate feedback on assessments or homework assignments.
- Coaches can respond to clients by linking them with experts from different information domains.
- Coaches can respond to clients by linking them with peers having similar challenges.

At Your Fingertips

- Assessment tools, such as 360-degree feedback and self-assessments, can be administered, at low cost, as needed.
- Client data can be collected and digitally stored.
- Tracking mechanisms support longitudinal assessments with comparative summative reports to measure progress.
- Coaches can provide client access to resources, including tools to help set goals and record and track progress.
- Digitally stored messages transform into valuable information for later reference.
- Sessions can be digitally stored, so clients can review goals, chart milestones, and review progress over time.
- Coaches can analyze session data to identify trends from their electronic communications.
- Clients can access electronic support systems, such as knowledge bases and data repositories, as needed.

- Coaches can direct clients to appropriate on-line learning assignments to prepare for coaching sessions.
- Coaches can review homework assignments on-line and provide feedback in preparation for coaching sessions.
- Technology makes possible the ability to track individual metrics of engagement, use, and satisfaction.
- Technology also provides that ability to track organizational metrics by quantitatively measuring and aggregating data to supply information to productivity improvement and return on investment studies.
- Organizations can also analyze the aggregate information to identify organizational strengths, gaps, and trends.

Despite these potential benefits, we are not advocating that e-coaching replace face-to-face coaching. In fact, we acknowledge the potential limitations of e-coaching and address these concerns later. As with most leadership development initiatives, there are advantages and disadvantages with their various modes of delivery. Therefore, the issue is not whether or not technology should be used as a mode for leadership coaching. Indeed, the practice is occurring and is likely to expand to support the growing coaching demand and needs of future leaders. Instead we propose that the issue is not *if* we should integrate technology into leadership coaching, but rather to determine *how* we can most effectively support coaching using technology.

WHAT DO WE KNOW ABOUT E-COACHING?

Leadership coaching, despite its popularity, has only relatively recently (1990's; see Kampa-Kokesch & Anderson, 2001) become an accepted mode of positive leadership development. As a result, coaching practices are often based on experiences and lessons published by successful coaching organizations and coaches (e.g., Center for Creative Leadership, Personnel Decisions International, Dr George Hollenbeck), as little empirical research has been conducted on coaching program practices and processes (Feldman, & Lankau, 2005; Peterson & Kraiger, 2003; Wilkins, 2000). So from an applied perspective, we know more about the practice of leadership coaching, such as what it is, why it's used, and that it is effective (Fisher, 2001; Peterson, 2002). However, from a theoretical perspective, we understand less about why leadership coaching is effective and what strategies can be effectively translated into a technology-supported environment.

Results of recent surveys by Corbett, Corbett, and Colemon (2008) and Frazee (2008) provide insights specifically on e-coaching practices. Nearly 1,300 world-wide respondents indicated that 51% of coaching communication is not "in-person," instead relying on technology that allows coaching delivery without any personal contact. A less representative sample of 191 coaches reported that while only 8% of coaching was occurring entirely face-to-face, only 26% of coaching was primarily or entirely occurring at a distance with technology supported communication and resources. E-coaching was primarily used by internal coaches to provide assistance on specific tasks or assignments; accelerate individuals' time-to-competency; and improve training transfer for executives, women, and expatriate managers.

The limited e-coaching research available also suggests that e-coaching is effective.[1] For example, Goldsmith and Morgan (2004) concluded no significant differences between coaching face-to-face or by phone. Two other studies (Wang & Wentling, 2001; Wadsworth, 2001) demonstrated that e-coaching significantly impacted training transfer. Unfortunately, the empirical e-coaching literature mirrors the broader coaching research in that little is known about why e-coaching is effective or the factors critical to ensuring effective e-coaching practices.

WHAT ARE KEY ISSUES AND CONCERNS WITH E-COACHING?

As a result of the lack of scientific information to guide e-coaching, we have developed a systematic framework to identify factors that may affect e-coaching practices. Informed by the general coaching, e-coaching, e-counseling, e-learning, leadership development, training, and technology literature, we present the following framework and discussion to raise awareness of critical issues, highlight potential concerns, and provide guidance for incorporating technology to support the growing demand for leadership coaching.

Introduction of Leadership E-Coaching Framework

Systems theory provides us with a tool to examine complex social systems and a framework to describe and analyze a group of things that work in concert to produce results. To paraphrase Katz and Kahn (1978, pp. 18–22), the first problem in understanding a social system is its location and identification. How do we know with what we are dealing? What are the boundaries? Who are the individuals whose actions are of interest? What behaviors belong within the system? An input-process-output or I-P-O model provides us with a means of identifying such boundaries and reducing the complex into something more manageable. The three central components of an I-P-O model are the *input,* or the external factors that enter the system; the

Framework for Leadership Coaching in a Virtual Environment

Coaching Outcomes

Reactions
- Satisfaction
- Utility

Learning
- Knowledge
- Cognitive
- Affective

Behavioral Skill Demonstration

Organization Impact

Formative Processes

Coaching Process

Mechanics
- Number, Duration, Frequency, and Timeliness of Sessions
- Session Preparation & Follow-up

Program Content
- Contracting & Confidentiality
- Assessments
- Action Planning
- Evaluating Progress
- Transitioning

Relationship
- Building & Maintaining Rapport
- Establishing & Maintaining Trust
- Encouraging Commitment
- Promoting Collaboration

Tools/Techniques
- Active Listening
- Questioning
- Feedback

Coach & Coachee Characteristics

Coach

Readiness
- Coaching Philosophy
- Competencies
- Experience
Motivation
Personality

Coachee

Readiness
- Skill Needs
- Developmental Goals
- Prior Coaching
Experience
  o Satisfaction
  o Medium
  o Skill Developed
Motivation
Personality

Coach-Client Match

Behavioral Preferences
- Personality
- Work Style

Commonalities
- Personal
- Education & Work Background
- Interests

Coach Capabilities versus Client Needs

Lisa Boyce, United States Air Force Academy
Gina Hernez-Broome, Center for Creative Leadership
©2007

process, or the actions taken upon the input materials; and *output*, or the results of the processing.

As the purpose of our I-P-O model is to frame the issues relevant to supporting coaching in an e-environment, we also include an additional component that may moderate or impact the relationship between the input and process (I –> P). Therefore, the four components illustrated in the model are Coach and Coachee Characteristics, Coach-Client Match, Coaching Process, and Coaching Outcomes. These four components should help guide your thinking as you integrate technology or review your current e-coaching practices. In addition, we hope researchers will consider this framework to systematically investigate and empirically support the growth of e-coaching practices. Each of these components will now be described and discussed.

Coach and Coachee Characteristics

As depicted in the model, the coach and client bring a level of readiness, motivation, personality, and previous experiences to the coaching process. Anecdotal, practitioner-oriented literature provides advice on who makes an effective coach as well as who should be coached.

For example in our opening scenario, Nayah, the coach, was described as having ten years of coaching experience and appears open to integrating technology. Unfortunately, many of us have encountered senior coaches as well as organizational leaders who were less than receptive to incorporating technology, especially as a means of communication in a relationship-based industry. One highly respected coach with a counseling background and nearly 30 years of experience felt that he "needed to see the client's face" to effectively read and understand the conversation and conducting a session by e-mail was not even an option. We have even encountered an organization that insisted that every meeting between a coach and their leaders receiving coaching had to be face-to-face because of concerns with miscommunication and possible perceptions by the coachees that they might not be receiving the "full monty." Most discussions with coaches were not as extreme, but a level of discomfort was apparent with most. Many believed that at a minimum initial personal contact is needed to establish rapport and that face-to-face dynamic is impossible to recreate via e-mail.

These concerns are not irrational. The lay literature (e.g., Dixon, 2008; Goldsmith, Govindarajan, Kaye, & Vicere, 2002; Rossett & Marino, 2005; Triad, 2001) suggests that the characteristics and requirements of a successful e-coach will differ slightly from that of the traditional face-to-face coach, and those more comfortable in the traditional face-to-face coaching relationships will have to adapt to meet the needs of the younger, more technologically savvy global clients. Of the coach and client characteristics, the coach's readiness factors (coaching philosophy, competencies, and experience) and the client's readiness factors (skill needs and developmental goals) as well as motivation and prior coaching experience may be the most relevant to coaching in an e-environment.

As a personal coaching philosophy establishes, clarifies, and reflects who you are as a coach and your coaching practice, it can also determine if and to what extent you employ technology into your coaching engagements. This is a very individual choice, and each individual coach needs to evaluate their coaching philosophy. Only by asking yourself, *Why am I coaching? Who am I coaching?* and *What kind of coach do I want to be?* will you be able to determine not only your comfort level but desire and ability to e-coach. The advantages of re-evaluating your personal coaching philosophy include an opportunity to reflect on yourself, your beliefs, values, experiences, and KSAs (knowledge, skills, and abilities). You will consider your strengths and weaknesses, including KSAs, which may require upgrading. Further, the process will help you gain an understanding of your complimentary as well as accommodating coaching method of delivery (i.e., you may prefer face-to-face meetings to deliver assessment feedback but are comfortable providing that feedback via phone and WebEx if your client can commit to a 45-minute uninterrupted session). Finally, your coaching philosophy defines your optimal and acceptable clientele. This, among other insights, provides you with a better understanding of your clients' interest and ability to use technology, better equipping you to tailor your coaching to meet their needs.

The Society of Industrial/Organizational Psychologists (SIOP) and the International Coach Federation (ICF), two non-profit organizations that represent professional or business coaches, provide a list of core coach requirements and competencies, which resemble the competencies suggested by the literature (Brotman, Liberi, & Wasylyshn, 1998; Hall, Otazo, & Hollenbeck, 1999; Graham, Wedman, & Garvin-Kester, 1994): ability to express active listening, ability to create and raise the client's awareness, high standards of personal and professional ethics, and expertise in adult learning and leadership, to name but a few. SIOP also provides some guidance on client-focused factors (e.g., admits to having a performance deficiency, desires to improve performance, ability to make progress towards goals, basic learning skills). Of particular interest is the extent to which these characteristics need to be adapted to support e-coaching.

The most obvious coaching competency requirements of an e-coach are those associated with the logistics of working with technology. Of specific concern is the need to accommodate the client's preferred technology support and communication tools. In most situations, leaders seeking coaching do not have the time to learn a new technology, so e-coaches need to be able to adapt to technologies their clients are comfortable using. Therefore, the onus is on the coach to understand the limitations of the technology (e.g., capabilities, bandwidth, information security, etc.), manage the intricacies of the technology to maximize its effectiveness, and leverage its capabilities to advance the relationship. In addition, the coach should monitor technology advancements that might be appropriate to gradually introduce later in the coaching process. While perhaps an obvious competency requirement, maintaining competence with the technology at a breadth and depth required to effectively e-coach is both a dynamic and continuous requirement as the various technology tools in an e-coach's repertoire will evolve with industry innovations.

E-coaches will also need to be more adept at communicating their thoughts and emotions in writing. In addition to writing clearly, e-coaches may need to explain complex issues in simple terms, translate non-verbal messages virtually, and ask more questions, such as "Why the silence? Are you surprised by my response?" and "Are you frustrated?" Whyman, Santana, & Allen (2005) believe that coaches will need to learn effective online coaching principles and offer several specific techniques to support computer supported communications, such as:

- Consciously use language that invites further interactions (e.g., "I'm looking forward to receiving your update next Friday. Please let me know how I can be helpful").
- Keep response short to accommodate the tendency of people to skim the computer screen.
- Make the main point first, then add supporting information, which contradicts a common coaching technique of first giving the supporting information, then building to a conclusion
- Use capitals to emphasize a single word, and typographical devices, known as emoticons (e.g., :-),:@, =O, etc.), and symbols (*, !) to express emotion and emphasis.

While these guidelines may not be comprehensive, they clearly illustrate that an e-coach will need to understand and possess a unique skill set to effectively communicate with their clients online.

As an e-coach is more likely to encounter cross-cultural coaching opportunities, possessing cross-cultural coaching skills may be advantageous. In today's global environment, leaders are working with international leaders and organizations and in multinational settings. In addition to the increased probability of a current client needing skills and support to adapt to divergent cultural attitudes, beliefs, and behaviors, clients from across the globe may seek e-coaching to support their integration into your culture. Knowledge of cultural behavioral change models and intercultural communication are essential, as is the ability to respond to a client in an international setting. For example, a client experiencing an intercultural conflict, preparing for a short-notice international meeting as they board their flight, or a foreign manager planning an assignment in your home country.

Perhaps not exclusive to an e-coach, but Goldsmith, et al. (2002) suggest that future coaches will need to possess a greater database and provide access to information and resources to meet the dynamic and instantaneous needs of their high-maintenance clients. As a result, the e-coach may not need diverse expertise, but the coach will need to be able to effectively identify and direct clients to solutions and problem-solving resources, such as other experts, knowledge bases, and materials. Therefore, knowledge of and access to web networking communities (e.g., LinkedIn, Facebook, SIOP Locator, TrainingConsortium, etc.), research and resource sites (e.g., APA, businesslink, webinars, ebooks, etc.), and public and private search engines will be necessary to effectively support clients' developmental requirements.

Our final coach readiness characteristic is experience. Pulley (2007) talks about the "learning curve" associated with technology supported communication. While many of the coaches Pulley interviewed expressed initial apprehension and discomfort, many reported a positive experience after learning and adjusting to the technology. Previous research has shown that technology experience positively relates to technology appreciation (e.g., Czaja & Sharit, 1998; Ellis & Allaire, 1999; Melenhorst & Bouwhuis, 2004), including recognition of enhanced and extended communication capabilities (Melenhorst, Rogers, & Bouwhuis, 2006) and the ability to translate skills to other technologically complicated devices (Kang & Yoon, 2008). Both anecdotal and empirical evidence appears to support the spiraling nature of technology experience in that experience breeds a desire and the skill to use the technology, which in turn expands experience, appreciation, and ability. Entry into this spiral, particularly for more reticent older coaches, is most likely to occur by focusing on the benefits of technology use (Melenhorst et al, 2006).

Similar to the coach, clients require a level of readiness to be coached in an e-environment. In addition to being comfortable and competent using technology, clients must also be able to effectively communicate in writing to use computer-mediated discussion tools. In fact, e-coaching may be more effective for coachees whose written language skills are stronger than their verbal skills. While most developmental goals on the surface appear suitable for e-coaching, we can think of a few cases in which individuals working on their interpersonal or "people" skills were more appropriately coached in a face-to-face environment.

Clients' prior coaching experience is also particularly relevant. Reasonably, individuals expressing satisfaction with previous coaching engagements and who successfully met their developmental goals with technology-supported coaching are likely candidates for additional e-coaching. We also know from related research that previous experience with technology is related to communication disclosure, satisfaction, perceived knowledge gain, and behavioral improvements (Calvin, 2005; Carey, Wade, & Wolfe, 2008; Shelton, 2004). Don't despair, though, if your client has not had previous coaching or technology experience. Coaches who consider, in advance, how to introduce technology to novice clients can achieve such positive outcomes within the duration of a program.

Client motivation may be a more critical factor in determining if a client is suitable for e-coaching. Anecdotal evidence (e.g., coaching.com and Carlson Co.) indicates that motivated volunteers to e-coaching programs resulted in higher levels of personal and business impact. Highly motivated clients may be more committed and accountable for their development and as a result are potentially more reflective and responsive, which is perhaps more crucial in an e-coaching environment. For example, because of the distance, the e-coach lacks control of the actual coaching session in which the client might choose to multi-task instead of focusing on the developmental opportunity or in an asynchronous conversation, the client may even fail to check the internet site and stay connected to the discussion.

As each client is unique as are their situation and needs, it would be wise to assess each client's readiness and motivation for e-coaching prior to integrating technology into your coaching relationship. Example questions to screen potential e-clients include:

- Have you experienced e-coaching before? Did you enjoy it? Did you find it helpful? Are you interested in e-coaching?
- Are you comfortable using technology? What are your preferred modes of communication?
- What is your motivation for seeking leadership coaching? Are you experiencing changes at work (e.g., with organizational structure, reporting relationships, new responsibilities)? Are you volunteering or are you mandated by your office to participate in leadership coaching?
- Can you commit to distance coaching? Will your work and family demands allow you to focus on your leadership development? Do you have management, technology, and administrative support?
- What do you consider your strengths and weaknesses? Would you categorize the following as strengths or skills needing improvement: self-management, accountability, interpersonal relations, conflict management, giving and receiving feedback, written communication (written and oral)?

Ultimately, leadership coaches need to understand each client, their experiences, abilities, personalities, goals, and why they are seeking coaching in order to determine if e-coaching is appropriate for that individual.

We've presented several coach and client characteristic factors for you to consider if you are planning to integrate technology into your leadership coaching practice or program. While the trend is moving towards e-coaching, e-coaching is not appropriate for every coach or client. From the perspective of the coach's readiness, you would be wise to first consider your personal coaching philosophy or the philosophies of each of your coaches. Coaches with a core belief that relation-based processes require face-to-face interaction and have no desire or interest in incorporating technology are not likely candidates for e-coaching. On the other hand, coaches who are open to alternative modes of interaction, excited by technology, and experienced and competent in using the technology or eager to learn the skills required of an e-coach are more likely to be effective coaching in an e-environment. Likewise, clients need to be assessed to determine their interest, motivation, and readiness to participate in e-coaching.

At this point, we again emphasize the lack of empirical research regarding the impact of coach and client characteristics and the use of technology on coaching processes and outcomes. For example, perhaps personality also plays a crucial role in e-coaching success. Are more conscientious individuals or individuals with openness to new experiences more appropriate for e-coaching? We therefore encourage systematic research to support identifying coaches and clients who can effectively coach and be coached in a virtual environment.

COACH–CLIENT MATCH

In addition to understanding who should coach and be coached, the match or fit between coach and client is critical to the partnership and the success of coaching outcomes. While research is relatively non-existent regarding this issue as well, practitioners (e.g., Ting & Hart, 2004; Hollenbeck, 2002) provide guidance on possible factors to consider when matching coaches with clients—for example, commonalities in personal, educational, work backgrounds, and areas of interest; compatibility in behavioral preferences, such as personality and work styles; and coaches' capabilities relative to the clients' developmental needs. Of particular interest is the importance of these factors in pairing clients to coaches operating in a virtual environment.

In our vignette, Nayah and Julius were depicted as compatible with well-matched personalities and similar learning styles despite having little in common—a combination that appears to work in their distal relationship.

Have you ever had an experience or heard of a situation in which a coach and client just did not click and as a result dissolved the coaching contract? What about a coaching relationship that connected almost instantaneously and the client and coach maintained a professional or social relationship even after the coaching had ended?

Take, for example, one coach who shared an experience in which the client essentially fired him after the second session. This was a coaching situation in which there was not a good match between the coach's coaching philosophy and style, and the needs and expectations of the client. The coach had a very inquiry-based style and took the approach of asking a lot of probing and open-ended questions in an effort to get the client to self-reflect and collaborate in setting his developmental goals. The client was looking for a coach with a much more directive style, someone who would essentially lay out a developmental plan and tell him what he needed to do. The client's feedback was that the coach was too passive and noncommittal. He, the client, expressed his frustration and sense of urgency in needing "things to happen quickly."

Another example that a colleague cited as the reason behind one of her most positive coaching relationships involved a series of unsuccessful attempts at finding a common ground with a potential client. After several starts and stops in trying to find something in common (e.g., sports, hobbies, volunteer activities, children, vacation plans, etc.), the discussion became a bit stilted until the client off-handedly commented on the perceived slowness of time. The coach had recently been to the Clock Museum in Vienna and matched the time comment with something she had seen on display. Coincidentally, the client had just completed reading *Time's Pendulum: From Sundials to Atomic Clocks, the Fascinating History of Timekeeping and How Our Discoveries Changed the World.* The two launched into an enthusiastic discussion about time, their knowledge of the invention of clocks, and more personal attitudes regarding punctuality, which expanded into a values discussion. Interestingly, references to time and clocks continued to pepper their later discussions, and the coach was impeccable with timely communications with this particular client.

These examples are validated through interviews with seasoned individuals who pair coaches to clients and evaluate the coaching partnership. The match between client and coach may be even more critical for e-coaching since there are greater opportunities for miscommunication, reduced use of verbal and nonverbal feedback, and penchant for immediate support. Thus, the means to develop and maintain the relationship needs to be more transparent. Of particular importance is the need for immediate credibility to be established from the onset, often based on information that is not weighed as heavily in a face-to-face coaching situation. For example, e-coaches can establish initial credibility by sharing with their client that they have previous experience with the client's organization or knowledge and experience with the client's industry. The educational background of the coach as it relates to the client's needs is also relevant, including supporting cultural or language requirements.

Many coaching coordinators also emphasize the need to find commonalities between the coach and client to help develop and maintain rapport, which is more difficult in an e-environment. Unfortunately, the value of matching clients and coaches based on similarities or differences in personalities, work, or learning styles is unclear. While one philosophy is that similarities may enhance rapport, others expressed concern that matches based on similar personalities and styles may minimize the potential for client growth in contrast to working intimately with someone different from themselves.

In addition to appreciating the potential importance of supporting an e-coaching relationship by pairing clients to coaches to magnify coach's credibility, promote shared personal and professional commonalities, and maximize compatibility and developmental needs, coaching organizations should also consider and examine technology tools to support the matching process. Innovative technology that relies on data processing software can identify the acceptable and optimal client-coach partnership by providing fit scores for each possible client-coach pair. A process that manually can take 10 minutes per client or multiple hours for programs with a large cadre of coaches can occur in a matter of seconds using technology.

Limited research (e.g., Boyce, Jackson, & Neal, 2007) encourages the use of technology in matching clients and coaches with the caveat that the appropriate algorithm may be unique to different coaching organizations and clientele, and clients should always be provided with an opportunity to be re-matched. A recent American Management Association study (Thompson et al., 2008) reported that 65% of terminated coaching assignments were due to mismatch between coach and employee. With coaching occurring in modalities other than face-to-face sessions for nearly half of the survey respondents, failing to match clients and coaches is both a waste of time and money. Further research is needed to validate the three factors of compatibility, commonalities, and coach capabilities/client needs and their impact on the quality of the relationship and coaching outcomes. Organizations need to, however, consider such factors now and invest in technology or time to match clients with coaches to promote successful relationships in a virtual environment.

COACHING PROCESS

The coaching process was partitioned into four major sub-processes: mechanics or the logistics of the coaching session (e.g., number, duration, frequency, and timeliness or responsiveness of sessions; session prep and closure including preparing an agenda, completing homework, and documenting the meeting; program content (e.g., contracting, establishing confidentiality, assessments, action planning, evaluating progress, and transitioning); the relationship (e.g., building and maintaining rapport and trust, encouraging commitment, and promoting collaboration); and tools and techniques (e.g., active listening, questioning, and feedback).

The practitioner literature has ample advice on program design, including the logistics, content, and person- and task-focused processes for face-to-face coaching practices. For example, The Executive Coaching Forum (2004) published *The Executive Coaching Handbook* in response to "the need for professional guidelines and practices" (p. 1). While making a plea for rigorous research, the handbook relies heavily on "practiced wisdom" and, when available, supports procedural guidance with research. Similarly, CCL (Ting & Hart, 2004, p. 129, 131) offers sage advice based on their practices, which, when possible, is also supported by research. Our interest, of course, is how e-coaching can effectively apply or adapt these processes.

Returning to our opening scenario, several aspects of potential e-coaching processes were illustrated. For example, Nayah's ability to immediately respond to Julius's emergent need reflects the dynamic and rapid environment of today's leaders but also the advantages afforded by technology in supporting the mechanics of e-coaching. Similarly, technology was integral to supporting the mechanics and logistics of preparing and coordinating Julius's next coaching session. The use of technology in the e-coaching program content was demonstrated by Nayah's use of technology to track on-line assessments, review action plans, and evaluate performance trends. However, the e-coaching relationship, which is perhaps the most contentious aspect of losing the face-to-face interaction, is not illustrated in the vignette, nor the tools and techniques as these coaching processes are not only difficult to illustrate in a short vignette, but deserve focused attention because of the ambivalence associated with the coach-client interaction in an e-context. Therefore, we offer more extensive examples from actual coaching experiences to illustrate.

Rapport is the heart of the coaching relationship. In face-to-face coaching interactions, this may be established with little effort and is often felt as a sense of ease, warmth, and genuine interest between the coach and coachee. Coachees often describe it as "just the right chemistry." In an e-coaching environment, the coach must be more deliberate and diligent in ensuring that rapport is established. One positive example of rapport being established within an e-coaching session was provided by an e-coach who works very hard to discover common areas of interest with the coachee, whether it is a sports team, music, or a favorite author asking very specific questions about the coachee's non-work interests. Whenever this coach comes across a news article or some other piece of information or trivia regarding that common interest, he emails it to his coachee along with a quick inquiry as to how things are going. It's a way of showing, "Hey, I remember you're interested in this too and because I'm genuinely interested in you, I remembered." This conveys to the coachee that even though there is no face-to-face interaction, the coach has a sense of who he is and thinks about him outside of email correspondence.

The above is in strong contrast to a situation in which the coach told of the importance for a particular client, a woman, to do the "soccer talk" to build rapport. The coach noted that she, the coach, had wanted to keep the discussion purely professional and work-related and had missed this vital piece of insight until after the fact. As a result, the coaching engagement never materialized because the client ended the coaching. When asked, the client responded that the coach had not once referred to or asked about her children. Again, in a face-to-face interaction, this type of discussion may more naturally emerge. In e-coaching it is critical to be very deliberate in asking questions about the client's background and family and gauge the importance of those areas to the client by asking them very directly. You can then incorporate that information, as appropriate, into your interactions with the client as you work to establish rapport and create an environment of trust.

Similarly, the communication tools and techniques employed in coaching are unique to each situation. As will be discussed later, in face-to-face coaching sessions, a critical technique utilized by coaches is active listening, which ensures that the coach accurately understands the coachee's point of view. While many of the same steps are important regardless of the coaching medium, some become particularly important to emphasize in an e-coaching environment due to the potential for misinterpretation and misunderstanding that can result from the lack of nonverbal cues such as facial expressions and body language. For example, clarity is of the utmost importance, and you should double-check on any issue that is ambiguous or unclear to you and request clarification: "Let me see if I'm clear. Are you referring to...?" Paraphrasing is also necessary in this context because it helps to confirm your understanding and assures the coachee that you are tracking accurately. Given the increased potential for misunderstanding in e-coaching, this again becomes a vital skill to incorporate into your communications with the coachee. Because there is a lack of nonverbal cues, the e-coach should verbalize his or her own reactions more frequently and ask more questions such as: "Are you angry?" "Is this a surprise?" or, in the absence of a response, "What is your non-response about?"

The following example demonstrates an effective e-coaching exchange.

Client initiates, "The last few weeks have been very difficult. We had a major shutdown, which went well. This was followed by some continuing operating problems, which resulted in a lot of micromanagement within our organization. I did not handle this very well. I began to have doubts as to whether I could succeed or want to be department manger. Things have gotten better this week, and I have been able to return to some of the action items that I've been working on. I still have some doubts. . . ."

The coach responds, "I was truly sorry to hear about the difficulties over the last several weeks. Shut downs are usually difficult in my experience. The micromanagement doesn't help, as you've experienced. You didn't say what your behavior was that led you to believe you didn't handle the situation well. If you would like to share the specifics with me, I may be able to respond more specifically. How you handle difficult times is as much a part of your development towards manager or higher as taking positive actions you have to improve relationships. What about your behavior disappointed you? How could you recognize those triggers sooner and take different actions?

These examples highlight the unique interactions both across and between coach-client relationships, adding greater difficulty in identifying successful comprehensive strategies to employ in an e-environment. However, the uniqueness of each coaching relationship and situation doesn't preclude identifying and applying techniques that will enhance e-coaching processes including developing and maintaining relationships and communicating in most virtual coaching environments. Keep this in mind as we step through the Coaching Process components of our model.

First, when examining the mechanics or logistics of e-coaching, the two most frequent concerns are the lag time between contacts and the lack of preparation by clients for the coaching session. Dixon (2008) discussed major difficulties with long gaps between communication between client and coaches, including misunderstandings, delayed reaction, and duplication of effort. Other practitioners have shared frustrations with the lack of productivity with synchronous discussions when clients fail to complete their assignments or reflect on interim asynchronous conversations or feedback.

These two concerns highlight the potential difficulties of coaching virtually. Coaches should consider all aspects of the mechanics and logistics of e-coaching and develop a plan to minimize frustrations and avoid miscommunications. The following suggestions are examples of possible techniques coaches might employ to enhance the e-coaching process.

- Set logistical rules, including frequency for checking websites or for messages, maximum response time (e.g., return call within four hours), interpretation or stipulation of the length, depth, or need for acknowledgement to asynchronous communications, turnaround time for homework or feedback.

- Establish a rhythm regarding the number, duration, and frequency of asynchronous and synchronous communications.
- Schedule formal coaching sessions with conditions for cancellations.
- Communicate expectations regarding response time, which includes availability or office hours—a particularly salient issue when client and coach are in different time zones.
- Use technology (e.g., interactive scheduling tools and auto reminders) to coordinate calendars.
- Prior to session, provide an overview or structured agenda including the purpose of the session and the scheduled meeting time and contact method.
- Prior to sessions, provide clients with relevant preparation materials to enhance the coaching session.
- Close session by agreeing to a future time or event in which to reconnect.
- After sessions, provide follow-up materials to enhance the coaching process.
- Encourage clients to initiate contact with questions as they occur, rather than waiting until the next coaching session.

Second, many models are available regarding program content, but most contain the five core activities we've included in our model: Contracting and Confidentiality, Assessments, Action Planning or Goal Setting, Evaluating Progress, and Transitioning. The most frequent concern with the content aspect of e-coaching is with contracting and confidentiality. The potential for confusion regarding the role of the coach exists in traditional face-to-face coaching but appears to be exacerbated in an e-environment (e.g., Triad, 2001). The potential for confidentiality violations also escalates with technology-supported discussion, particularly when the clients use their organization's technology to communicate, as many organizations own and monitor their communication media (e.g., e-mail). Many forms of computer-mediated communication are also stored, which means the information can be later accessed. On the other hand, other e-coaching activities generally are perceived to benefit from the use of technology. For example, the stored written communication provides a tool for the coach to understand and provide feedback on trends and progress to the client, and from the coachee's perspective, the stored communication can serve as a source of information for the future.

Similar to our discussion regarding the mechanics and logistics of the e-coaching process, we believe that coaches should consider each component of their coaching program and develop a plan to clarify expectations and avoid breeches in confidentiality as well as maximize the advantages of integrating technology. The following suggestions illustrate issues to consider and guidelines coaches might employ to advance their e-coaching process.

Contacting and Confidentiality

- Discuss with the clients their expectations as well as the e-coaching process.
- Use technology to clarify expectations and present information (e.g., video, web testimonials).
- Clarify coaching roles initially and reinforce repeatedly if needed, including defining professional and personal boundaries.
- Create a mutual contract, review and if necessary update routinely.
- Clarify payment plan (e.g., by number of contact, time spent, flat rate over time).
- Minimize use of company technology; at most use company communication tools for administrative purposes only.
- Use technology to schedule and send auto-reminders of standard confidentiality conflicts.

Assessments

- Conduct on-line assessments that support rapid feedback.
- Conduct sensitive discussions on the phone or in person; not only are corporations required by law to keep records of e-mail, but they can be called into evidence in trials and can become public record.

Action Planning and Evaluating Progress

- Provide guidance and direct clients to resources (e.g., goal-setting tools, learning activities or exercises, etc.) and ensure resources are in usable format (e.g., e- or hard copy).
- Maintain client "paperwork" (e.g., assessments, goal statements, homework, communications etc.) on a mutually assessable site to support trend and progress evaluation.
- Between sessions, employ auto reminders (e.g., via e-mail, text messaging) to help client stay focused on the coaching process.
- Use all appropriate, available, and cost-effective communication to support progress toward goals.

Transitioning

- Celebrate achievements virtually.
- Develop, discuss, and agree on an exit strategy that supports transitioning and closure virtually.

Third, as highlighted earlier, the client-coach relationship is an integral aspect of the coaching process. In fact, a quality coaching relationship is often considered the single-most important factor for successful coaching outcomes (e.g., Assay & Lambert, 1999; Gyllensten & Palmer, 2007; Wasylyshn, 2003). The four key processes associated with the client-coach relationship are building and maintaining rapport, trust, commitment, and collaboration (Ely et al., in press). These four social constructs involve a mutual responsibility between a coach and client and as a result may be difficult to develop in a coaching relationship as the coach cannot accomplish the process alone. Further, the non-verbal behaviors associated with these relationship-based processes are considered critical to achieving and assessing effective attainment (Dimatteo & Taranta, 1979; Bernieri, Gillis, Davis, & Grahe, 1996), which creates an additional burden in a virtual environment (Hian, Chuan, Trevor, & Detenber, 2004).

While a series of research studies by Walther (e.g., Walther & Burgoon, 1992; Walther, 1993, 1995, and 1996) suggests that relationship development, in general, is equivalent in computer-mediated and face-to-face communications, the anecdotal evidence suggests that the process is not so straightforward. Therefore, coaches would be wise to give careful consideration to relationship processes and actively employ strategies to promote partnerships in a virtual environment. Example recommendations include the following:

Build and Maintain Rapport

- Initiate disclosure and sharing professional and personal information to quickly identify commonalities and develop rapport.
- Incorporate the commonalities, feelings, and comments from prior sessions.
- Use people's names more than you normally would face-to-face.
- Demonstrate you are paying attention by asking open-ended questions and rephrasing.
- Identify shared experiences and discuss similar and dissimilar perceptions of those experiences.
- Mirror communication (for example, match client's preferred level of information; preferences for visual, auditory, kinesthetic learning modalities; and key words and phrases).
- Schedule the first two to four sessions close together to establish a foundation and build trust.

Establish and Maintain Trust

- Establish a safe and supportive environment by developing a personal connection.
- Identify a trust link (an individual that you both know and trust).
- Establish communication rules to increase awareness and clarify miscommunication, emotions, or misunderstandings.

- Provide substantive and timely responses.
- Follow up on deliverables and explain reasons for any delays.
- Address perceived discontent as early as noticed.
- Review unsatisfactory dialogue.

Encourage Commitment

- Encourage commitment by identifying in writing goals and milestones.
- Establish clear definition of responsibilities.
- Hold each other accountable by providing notes and responsibilities after sessions.
- Identify external influences that will impede goal achievement and employ tools to minimize.
- Ensure a regular pattern of communication.
- Involve client in meaningful tasks as soon as possible.
- Highlight early successes.
- Share and have your client share the milestones accomplished of an overall goal.
- Celebrate successes (e.g., send a virtual card, schedule a virtual celebration, or share the success with client's organization).

Promote Collaboration

- Promote collaboration by openly discussing the need to collaborate.
- Openly share information and resources in moderation to reduce information overload.
- Develop an atmosphere of learning by demonstrating a learner-centered focus and responding to different learning styles.
- Allow time for thoughts and reflection to incubate.
- Employ collaboration tools (e.g., Sharepoint, GoogleApps, LinkUAll, InfoWorkspace, Lotus Notes, WebEx, etc.); do not rely solely on e-mail.
- Establish a completion ritual at the end of each session to build emotionally connected relationship.

Technological approaches to coaching have a reputation for being impersonal and many believe that "much of the emotion is left out" making it difficult to build rapport and establish a trusting relationship (Rossett & Marino, 2005). However, a survey of CCL Leadership Development Program participants, who utilized follow on web based coaching, indicated that their coaches gave a human touch to the technology and provided a way to reach out and connect. To further illustrate these successful e-coaching interactions, we provide two examples from these CCL e-coaches who incorporate many of the recommendations listed above.

> One e-coach wrote, "With both of the goals you've outlined I would really welcome and enjoy the opportunity to virtually address them, brainstorm and consider, and offer you feedback. I would also ask that you take stock of your many talents and use these in support of your goals. For the next several weeks, to paraphrase you, your battles are my battles, and I am eager to be at your side as you move forward."

> Another e-coach wrote, "Thanks for the update, and for your focus on very important goals. I hope that you are finding your progress rewarding and finding that the effort and initiative it takes to get where you are is not too taxing. Although you can't see it directly, I am providing you with a lot of cheering and cheer-leading from afar!"

Fourth, coaches employ many tools (e.g., contracts, bio-measures, personality assessments, goal planning and setting forms) to support leadership development, and to a large extent these tools can easily be translated into an e-environment. However, the techniques used by coaches focusing primarily on communication (e.g., active learning, questioning, and providing feedback) present a greater challenge virtually. Face to face, these techniques often rely on non-verbal skills (e.g., eye contact, leaning forward, nodding, sincere expression of emotion, etc.), which are often difficult to demonstrate online.

For example, borrowing from the counseling research, we are aware that e-counseling may be less effective for in-depth processing (Mallens, Day, & Green, 2003). When compared to face-to-face relationships, the authors hypothesized that the lack of contextual cues, less emotional arousal, and lack of on-line discussion experience may have impacted the effectiveness of confrontational interventions. Translate that research to coaching and you can clearly see why coaches must consider how to effectively provide critical feedback to their coachees. Whyman and her colleagues provide specialized training in techniques for on-line support processes and include specific guidance in providing client feedback, such as "make the main point first and then add supporting information; look for behavioural patterns, and make them explicit; and use capitals to emphasize a singe word" (Whyman, Santana, & Allen, 2005, p. 16).

The following guidance, which is by no means comprehensive, is put forth for coaches and coaching organizations to provide examples of how to reflect on issues that need to be considered when employing virtual tools and techniques. The bottom line is to ensure that tools and techniques employed in e-coaching accomplish their intended purposes, whether they are the same, modified, or an original effort.

Active Listening:

- Affirm coachees' feelings about working on their goals, and recognize how others, such as their colleagues, family, members, and friends, may be reacting.
- Summarize by rephrasing with an emphasis on emotions and facts.
- For example, an effective e-coach wrote, "I'm glad you like the idea having people observe you in meetings and provide direct feedback. That should be productive to work on when you return from vacation."

Questioning:

- Ask questions and offer suggestions rather than being directive.
- Have the coachee look toward a future direction with a specific call to action.
- Create options and describe the benefit of the change.
- Integrate as appropriate with other resources or assessments.
- For example, another e-coach asked, "For those who are new to working with you, have you considered sharing specifically with them that you are experimenting with some new behaviors?"

Feedback:

- Recognize and reinforce coachees' progress toward their goals and acknowledge that it takes hard work.
- Focus on behavioral goals and reinforce coachees' progress towards attaining those goals through their action steps.
- Balance critical feedback with appropriate praise.
- For example, in one session an e-coach wrote, "I have found that true progress happens in varying ways. There will be bumps in the road, as you have already experienced. The best part about it is that you've worked hard to stay present and realized what was happening. The more you do this the more often you have success."

While these are examples of good coaching techniques in general, when applied to e-coaching it is particularly critical to be deliberate and conscious about the language used, making sure that responses are substantive and specific. Further, these comments become available for review and reflection as well as for compilation, a tool both the coach and coachee can use to identify trends, patterns, and movement. Finally, when applied effectively, a digital tool or technique can facilitate thoughtful communications with a medium that allows the coach and coachee to reflect and clarify their thoughts before corresponding with each other.

The coaching processes, including the mechanics or logistics, program content, building and maintaining coach-client relationships, and coaching tools and techniques, were discussed with a focus on the unique challenges presented with coaching in a virtual environment. Similar to our previous comments regarding the lack of research, systematic examination of these process factors is needed to understand and effectively perform e-coaching. Based on the limited research and anecdotal experiences, we've offered guidance to support your thinking about the process of actually coaching in an e-environment. Our intent is that the recommendations and examples help you and your coaching organization reflect on the issues that present challenges and develop plans to manage them.

COACHING OUTCOMES

Leadership coaching is qualitatively different from most approaches to leadership development and therefore holds particular challenges for evaluating coaching outcomes (Ely et al., in press). Adding technology to the already unique nature of a one-on-one coaching process increases the inaptness inherent to only applying traditional training evaluation to coaching. Ely and her colleagues addressed the general difficulties of evaluating coaching by applying a systematic framework to the unique aspects of our conceptual model of leadership coaching (Figure 8.1).

An integrative evaluation framework captures both the summative and formative needs by adapting a common training evaluation model (Kirkpatrick, 1976; 1994) to reflect the nuances of leadership coaching. A highly accepted framework for categorizing training outcomes is Kirkpatrick's four-level typology, which serves as a useful foundation for identifying relevant leadership coaching outcomes. The four criteria are reactions, learning, behavior, and results. For those less familiar with Kirkpatrick's work, we will provide a brief summary of the four criteria and our extension of those criteria to support the uniqueness of coaching. We will then discuss these summative and formative outcomes with a focus on the evaluation opportunities afforded the e-coach.

Reactions refer to the subjective evaluations individuals make about their training experience. Other typologies (e.g., Alliger, Tannenbaum, Bennet, Traver, & Shotland, 1997) suggest that perceptions of coaching (e.g., general satisfaction with the coach, the coaching relationship or match, and the coaching program itself) are multidimensional and include both a satisfaction and utility judgment. Utility judgments are the belief about the value and usefulness of the coaching program, the extent to which they believe their involvement in coaching has positively impacted their job performance or ability to perform their job.

Learning captures what participants have learned from engaging in the coaching process. Immediate knowledge is the assessment of knowledge acquisition at the conclusion of the coaching program. Knowledge retention is the assessment of the retention of knowledge at some point after the immediate conclusion of the coaching program. These concepts are combined into a single knowledge criterion. As knowledge outcomes are typically declarative and procedural and learning is also multidimensional, we've extended the category to include cognitive and affective learning outcomes (Kraiger, Ford, & Salas, 1993). Cognitive learning includes two outcomes particular to coaching: self-awareness and cognitive flexibility. Affective learning is the change in attitudes, motivation, and goals relevant to the coaching objectives.

Behavior refers to the influence of coaching on leadership or job-related behaviors, the actual demonstration of behavioral skills to the work setting or learning transfer. Leadership performance is specifically included as that is the primary purpose of leadership coaching. In assessing outcomes, leadership behaviours can be classified into four broad dimensions: information search and structuring, information use in problem solving, managing personnel resources, and managing material resources (Fleishman, Mumford, Zaccaro, Levin, Korotkin, & Hein, 1991).

Finally, organizational impact is the objective tangible results, such as reduced costs, improved quality or quantity, and revenue. Since leadership involves influencing others, evaluating organizational outcomes must include impact on subordinates, peers, and superiors. For example, subordinate-level variables might include job satisfaction, retention, and work group productivity.

Formative outcomes focus on process criteria, as opposed to the summative criteria, and provide additional information to understand and improve coaching interventions. Formative evaluation concentrates on the coachee, coach, and coaching process. For example, assessing the relationship factors (rapport, trust, commitment, and collaboration) provides valuable insight on the client–coach relationship. Assessing formative criteria periodically can help prevent a relationship from derailing as well as provide data on why coaching program outcomes and their coaches are effective or ineffective.

Combined, these summative and formative criteria provide a framework for assessing the effectiveness of a coaching program, including coach and coachee readiness, coach-client match, and coaching process—particularly, the coaching relationship. While researchers and practitioners appear to be evaluating summative outcomes, primarily leaders self-reporting their behavioural development, in face-to-face coaching (Ely et al., 2008), the evidence of evaluation for e-coaching is not equally available. With online coaching gaining popularity, it is essential that researchers and practitioners make greater efforts to conduct and report the findings of summative and formative outcome evaluations of leadership e-coaching. Empirical confirmation of e-coaching effectiveness and insight on how technology can be effectively employed is necessary to promote evidenced-based practice.

The few examples of reported e-coaching outcomes are encouraging but highlight the value of formative evaluation to understanding how to effectively employ technology. For example, researchers Wang and Wentling (2001) concluded that e-coaching had both positive and negative impact on behavioral skills demonstration. However, their discussion regarding e-coaching activities that significantly impacted learning transfer (providing resources, building relationship, building commitment, and supporting technology communication; but not feedback) allows critical reflection on why certain activities did and did not have positive impact. Similarly, practitioner Ken Blanchard's (2008) coaching.com website provides testimonials from clients touting reduced costs, improved sales, increased productivity, and improved employee retention. However, the description of their web-based program (emotional learning component, online pre- and post- 360-degreed assessments, performance support tools, action planning with supervisors) provided serviceable information.

Evaluating coaching outcomes is obviously important to documenting the effects of a coaching program as well as to understanding the issues and factors that are critical to its development and implementation. In addition, this information is important to improving your own coaching practice; it can offer evidence for others seeking ideas and strategies for delivering e-coaching successfully, as well as providing reassurance to stakeholders investing in e-coaching. Thus, it would be prudent to consider issues relevant to assessing e-coaching outcomes. Following are examples of opportunities technology affords the virtual coach for the purpose of evaluation. In general, the e-environment provides an opportunity to collect a wide range of data for each coaching outcome that can be tracked and analyzed over time.

Reactions

- Collect satisfaction and utility data anonymously through outsourced survey programs (e.g., Survey Monkey).
- Eliminate the time and manpower required to hand enter data from hard copies by using survey software over the web.

Learning

- Incorporate automated graphing features to review goals, chart milestones, and review progress over time.
- Imbed measures in learning tools to assess pre- post- training tool effectiveness.

Behavioral

- Collect assessment data (e.g., 360-degree tools) to provide baseline data as well as measuring change.
- Use digital diaries to prompt the coachee to encode behavioral demonstration on the job.

Results

- Collect quantitative data that can be aggregated to assess organizational benefits of coaching.
- If multiple people from an organization are in a coaching program, organizations can use the aggregate data to track and measure results.

Formative

- Integrate embedded metrics on engagement, tool use, and the coaching process (e.g., time-spent on homework; Sharepoints "check-out" process identifies documents used, response time, duration of sessions, ratio of synchronous to asynchronous communication).
- Compile and analyze communication patterns (e.g., ratio of coach to coachee words, use of active listening and questioning techniques, rapport maintenance dialogue).

Leadership coaching in general, and e-coaching in particular are qualitatively different from traditional approaches to training. As a result, the criteria associated with evaluating training need to be adapted to support coaching outcomes. While conditions may be more challenging an e-environment (e.g., identifying stakeholders, acceptable survey return rates, assessing the different types of technology, etc.), technology provides advanced capabilities as well as conveniences in collecting and analyzing both summative and formative data. Practitioners and researchers must be careful not to abuse the technology by carefully identifying the critical criteria and planning a systematic evaluation system that will benefit not only the coaching program but coach, coachee, and coaching stakeholders.

Consideration of the Leadership Coaching Framework

Employing an I-P-O model, we've presented a leadership coaching framework to help support the consideration of key issues to coaching in a virtual environment. The four components of the Leadership Coaching Framework included Coach and Coachee Characteristics, Coach-Client Match, Coaching Process, and Coaching Outcomes. For each component we provided an overview of the issue, shared real-life examples and results of research, highlighted concerns, and offered guidance and recommendations for further consideration.

While many of the issues of leadership coaching are also relevant to coaching in an e-environment, the addition of technology presents challenges that need to be considered by those planning to coach virtually, currently using technology, or evaluating the effectiveness of their e-coaching programs. We've attempted to focus our discussions on those unique challenges, including at every opportunity to emphasize the need for systematic research to support the practice of e-coaching. Here again, we'd like to emphasize that while we are research practitioners, and despite our desire to formulate opinions based primarily on research, many of our suggestions and points of guidance are based on experience, knowledge of related fields, and anecdotal evidence, as, much to our chagrin, empirical research is just not available.

That said, as with every effort, this chapter is not without additional limitations. For instance, we have not discussed at any length the limitations often experienced by e-coaching, which include many of the same limitations as traditional coaching but are likely compounded as e-coaching often occurs over distances (e.g., lack of support from the organization, transferring learning into workplace behaviors, managing multiple stakeholders). Also, we did not include a discussion of blended coaching (combining both face-to-face and virtual coaching), which may moderate many of the specific limitations resulting from the use of technology that we have discussed (e.g., access and ability to use technology, building online relationships, employing techniques virtually) while maintaining the benefits of e-coaching. In addition, we often provided summary overviews of issues that may warrant a full chapter in themselves (e.g., the coaching relationship, coaching evaluation), and we encourage you to examine the cited references to gain a deeper grasp of these factors, which can only serve to better equip you as an e-coach or program manager.

Finally, the fundamental question of this chapter and our framework is, "How do *we* effectively leverage technology and online environments to support virtual coaching?," with the *we* referring to managers of coaching programs, the coaches themselves, and academicians. For example, technology is an excellent equalizer. Current concerns with coaching credentials, experience, and quality of service provided can be tracked, assessed, and supplemented with the aid of technology. Training can target coaches with ineffective online communication skills, and technology allows availability of support tools to all coaches, not just those with convenient access to a "coaching library" and years of "networked" resources. Technology supports the collection and advertisement of common outcome measures permitting comparisons across coaching organizations and between organization's coaches. Technology can support coaches' assessment of their own competencies and an understanding of the commonalities and compatibilities (or lack thereof) with clients, which can support real-time adjustment of their approach accordingly. Coaches can monitor and review conversations to assess the state of the relationship and adapt their strategy throughout the coaching relationship. Academicians can examine qualitative and quantitative data stored in the digital intake forms, assessments, conversations, and outcome surveys to build an understanding of the interrelationships of the four components of the Leadership Coaching Framework. The potential implications of e-coaching are limited only by our resourcefulness. We encourage you to actively consider both the potential and the limitations of integrating technology into e-coaching and extrapolate these concepts into e-mentoring.

THE FUTURE OF E-COACHING

The future will bring more e-coaching. Technological connectivity *is* transforming the way people live and interact. More transformational than the technology itself is the shift in behavior that it enables. We work not just globally but instantaneously. We are forming communities and relationships in new ways (The McKinsey Quarterly, 2006). This, in combination with the increased popularity of leadership coaching makes the promising future of e-coaching undeniable.

Leadership coaching is an integral component of leadership development programs, with nearly 70 percent of organizations with a leadership development strategy employing coaching. E-coaching is a tool of particular relevance for developing leaders in a dynamic, overscheduled, global world. All things considered, it's evident that technology greatly eases the process of making coaching scalable throughout an organization as well as throughout the world. The key is making coaching initiatives accessible to a broad audience of users without compromising any of the characteristics or dynamics that make coaching work, particularly that of the relationship between a client and a coach. The future success of e-coaching balances on integrating technology thoughtfully and systematically.

NOTE

1. For a summary of empirical research examining the effectiveness of face-to-face leadership coaching see Ely et al. (2008) and Passmore & Gibbes (2007).

REFERENCES

Alliger, G. M., Tannenbaum, S. I., Bennett, W., Jr., Traver, H., & Shotland, A. (1997). A meta-analysis of the relations among training criteria. *Personnel Psychology, 50*, 341–358.

Assay, T. P., & Lambert, M. J. (1999). The empirical case for the common factors in therapy: Quantitative findings. In C. Sills (Ed.), *Towards the coaching relationship* (pp 32–34). Reprinted in Training Magazine (Feb, 2003).

Bernieri, F. J., Gillis, J. S., Davis, J. M., & Grahe, J. E. (1996). Dyad rapport and the accuracy of its judgment across situations: A lens model analysis. *Journal of Personality and Social Psychology, 71*(1), 110–129.

Blanchard, K., (2008). *Client results.* Retrieved June 21, 2008 from the Ken Blanchard Companies The Leadership Difference Web site: http://www.kenblanchard.com/results/default.asp

Bolch, M. (2001). Proactive coaching. *Training, 38*(5), 58–63.;

Boyce, L. A., Jackson, R. J., & Neal, L. (2007). A Case for E-Matching. In L. A. Boyce & G. Hernez-Broome (Chair), *E-coaching: Supporting leadership coaching with technology.* Symposium conducted at the 22nd Annual Conference of the Society for Industrial and Organizational Psychology, New York, NY.

Boyce, L. A. & Hernez-Broome, G. (2007). Introduction. In L. A. Boyce & G. Hernez-Broome (Chair), *E-coaching: Supporting leadership coaching with technology.* Symposium conducted at the 22nd Annual Conference of the Society for Industrial and Organizational Psychology, New York, NY.

Boyce, L. A., & Ritter, A. (2002). *Executive coaching: The professional personal trainer.* Retrieved January 7, 2002 from Society of Industrial and Organizational Psychology Web Site: http://www.siop.org/Media/News/NewsReleases.htm

Brotman, L. E., Liberi, W. P. & Wasylyshn, K. M. (1998). Executive coaching: The need *for standards of competence. Consulting Psychology Journal: Practice and Research, 50*(1), 40–46.

Calvin, J. (2005). Explaining learner satisfaction with perceived knowledge gained in Web-based courses through course structure and learner autonomy. *Dissertation Abstracts International Section A: Humanities and Social Sciences, 66*(5–1), 1632.

Carey, J. C., Wade, S. L. & Wolfe, C. R. (2008). Lessons learned: The effect of prior technology use on web-based interventions. *CyberPsychology & Behavior, 11*(2), 188–195.

Corbett, B., Corbett, K., & Colemon, J. (2008). *The 2008 Sherpa executive coaching survey.* West Cheter, OH: Author.

Criswell, C. & Martin, A. (2007). 10 trends: A study of senior executives' views on the future. *A CCL Research White Paper.* Greensboro, NC: The Center for Creative Leadership.

Czaja, S. J., & Sharit, J. (1998). Age difference in attitudes towards computers. *Journal of Gerontology, 53,* 329–340.

Dimatteo, M. R., & Taranta, A. (1979). Nonverbal communication and physician-patient rapport: An empirical study. *Professional Psychology, 8,* 540–547.

Dixon, K. (2008). *Ecoaching: How it works.* Retrieved April 21, 2008, from Academme Learning Solutions Web site: http//www2.academee.com/html/news-case/articles/article-3.html

Elder, E., & Skinner, M. (2002). Managing executive coaching consultants effectively. *Employment Relations Today, 29*(2), 1–8.

Ely, K., Boyce, L. A., Nelson, J. K., Zaccaro, S. J., Hernez-Broome, G., & Whyman, W. (in press). Evaluating leadership coaching: A review and integrated framework. *The Leadership Quarterly.*

Ely, K., Nelson, J., Boyce, L. A., & Zaccaro, S. (2008). Evaluation methodologies of leadership coaching. In G. Hernez-Broome & L. A. Boyce & (Chair), *Leadership coaching eEffectiveness: Incorporating evaluation methodologies in practice and research.* Symposium conducted at the 23rd Annual Conference of the Society for Industrial and Organizational Psychology, San Francisco, CA.

Ellis, R. D., & Allaire, J. C. (1999). Modeling computer interest in older adults: The role of age, education, computer knowledge, and computer anxiety. *Human Factors, 41,* 345–355.

The Executive Coaching Forum. (2004). *The executive coaching handbook: Principles and guidelines for a successful coaching partnership* (3rd. ed.). Retrieved February 22, 2006 from http://www.theexecutivecoachingforum.com/handbook2.htm.

Feldman, D. C., & Lankau, M. J. (2005). Executive coaching: A review and agenda for future research. *Journal of Management, 31*(6), pp. 829–848.

Fisher, A. (2001). Executive coaching: With returns a CFO could love. *Fortune, February 19,* p. 250.

Fleishman, E. A., Mumford, M. D., Zaccaro, S. J., Levin, K. Y., Korotkin, A. L., & Hein, M. B. (1991). Taxonomic efforts in the description of leader behavior: A synthesis and functional interpretation. *Leadership Quarterly, 2,* 245–287.

Frazee, R. V. (2008). *E-coaching in organizations: Mapping the terrain.* Presentation conducted at the 2008 EDTEC Alumni Conference, San Diego, CA.

Goldsmith, M., Govindarajan, V., Kaye, B., & Vicere, A. A. (2002). *The many facets of leadership.* New York, NY: Financial Times Prentice Hall.

Goldsmith, M., & Morgan, H. (2004). Leadership is a contact sport: The "follow-up factor" in management development. *Strategy + Business, 36,* 71–79.

Graham, S., Wedman, J. F., & Garvin-Kester, B. (1994). Manager coaching skills: What makes a good coach. *Performance Improvement Quarterly, 7*(2), 81–97.

Gyllensten, K., & Palmer, S. (2007). The coaching relationship: An interpretative phenomenological analysis. *International Coaching Psychology Review, 2*(2), 168–176.

Hall, D. T., Otazo, K. L., & Hollenbeck, G. P. (1999). Behind the closed doors: What really happens in executive coaching. *Organizational Dynamics, 27*(3), 39–53.

Hernez-Broome, G., Boyce, L. A., Whyman, W. (2007). Critical Issues of Coaching with Technology. In L. A. Boyce & G. Hernez-Broome (Chair), *E-coaching: Supporting leadership coaching with technology.* Symposium conducted at the 22nd Annual Conference of the Society for Industrial and Organizational Psychology, New York, NY.

Hian, L. B., Chuan, S. L., Trevor, T. M. K., & Detenber, B. H. (2004). Getting to know you: Exploring the development of relational intimacy in computer-mediated communication. *Journal of Computer Mediated Communication, 9*(3), 1–24.

Hollenbeck, G. P. (2002). Coaching executives: Individual leader development. In R. Silzer (Ed.), *The 21st century eExecutive: Innovative practices for building leadership at the top* (pp. 137–167). San Francisco, CA: Jossey-Bass.

Howard, A. (1995). A framework for work change. In A. Howard (Ed.), *The changing nature of work* (pp. 3–44). San Francisco: CA: Jossey-Bass.

HR Focus (2008). *What is the 'most important' metric? 85*(2), 13–15.

Johnson, H. (2004). The ins and outs of executive coaching. *Training, 41*(5), 36–41.

Kampa-Kokesch, S., & Anderson, M. Z. (2001). Executive coaching: A comprehensive review of the literature. *Consulting Psychology Journal: Practice and Research,* *53*(4), 205–228.

Kang, N. E., & Yoon, W. Ch. (2008). Age- and experience-related user behavior differences in the use of complicated electronic devices. *International Journal of Human-Computer Studies, 66*(6), 425–437.

Katz, D., & Kahn, R. L. (1978). *The social psychology of organizations.* New York: Wiley & Sons.

Kirkpatrick, D. L. (1976). Evaluation of training. In R. L. Craig (Ed.), *Training and development handbook: A guide to human resource development* (2nd ed., pp. 1–27). New York: McGraw Hill.

Kirkpatrick, D. L. (1994). *Evaluating training programs: The four levels.* San Francisco: Berrett-Koehler.

Kraiger, K., Ford, J. K., & Salas, E. (1993). Application of cognitive, skill-based, and affective theories of learning outcomes to new methods of training evaluation. *Journal of Applied Psychology, 78,* 311–328.

Mallens, M. J., Day, S. X., & Green, M. A. (2003). Online versus face-to-face conversations: An examination of relational and discourse variables. *Psychotherapy: Theory, Research, Practice, Training, 40*(1/2), 155–163.

Marino, G. (2005). *E-Coaching: Connecting learners to solutions.* Retrieved June 21, 2008 from San Diego State University Web Site: http://edweb.sdsu.edu/people/ARossett/pie/Interventions/ecoaching_2.htm

Melenhorst, A. S., & Bouwhuis, D. G. (2004). When do older adults consider the Internet? An exploratory study of benefit perception. *Genontechnology, 3,* 89–101.

Melenhorst. A. S., & Rogers, W.A., & Bouwhuis, D. G. (2006). Older adults' motivated choice for technological innovation: Evidence for benefit-driven selectivity. *Psychology and Aging, 21*(1), 190–195.

Olson, M. L. (2001). *E-Coaching.* Retrieved April 21, 2008, from Learning Circuits: American Society for Training and Development for E-Learning Web Site: http://www.learningcircuits.org/2001/sep2001/olson.html.

Passmore, J., & Gibbes, C. (2007). The state of executive coaching research: What does the current literature tell us and what's next for coaching research. *International Coaching Psychology Review, 2*(2), 116–129.

Peterson, D. B. (2002). Management development: Coaching and mentoring programs. In K. Kraiger (Ed.) *Creating, implementing, and managing effective training and development* (pp. 160–191). San Francisco: Jossey-Bass.

Peterson, D. P. & Kraiger, K. (2003). A practical guide to evaluating coaching: Translating state-of-the art techniques to the real world. In J. E. Edwards, J. C. Scott, & N. S. Raju (Eds.) *The Human Resources Program-Evaluation Handbook* (pp. 262–282). Thousand Oaks, CA: Sage.

Pulley, M. L. (2007). Blended Coaching. In L. A. Boyce & G. Hernez-Broome (Chair), *E-coaching: Supporting leadership coaching with technology.* Symposium conducted at the 22nd Annual Conference of the Society for Industrial and Organizational Psychology, New York, NY.

Rossett, A. & Marino, G. (2005). If coaching is good, the e-coaching is.... *Training and Development (T + D), November,* 46–49.

Shelton, J. A. (2004). Technology-mediated interactions: An exploration of the effects of previous technology experience and comfort on self-disclosure in virtual environments. *Paper presented for the PsiChi Research Grant 2003–2004 Winners.* Retrieved August 17, 2008, from http://www.psichi.org/awards/winners/summer_03_04grant.asp

The McKinsey Quarterly (July, 2006). *Ten trends to watch in 2006.* Retrieved August 1 2006, from http://www.mckinseyquarterly.com/article_page.aspx.

Thompson, H. B., Bear, D. J., Dennis, D. J., Vickers, M., London, J., & Morrision, C. L. (2008). Coaching: A global study of successful practices. Retrieved June 29, 2008 from American Management Association Web Site: http://www.amanet.org/research/pdfs/i4cp-coaching.pdf

Ting, S., & Hart, E. W. (2004). Formal Coaching. In C. D. McCauley and E. Van Velsor (Eds.) *The Center for Creative Leadership Handbook of Leadership Development* (pp. 116–150). San Francisco: Jossey-Bass.

Triad (2001, October). Executive summary impact evaluation on the Coaching.com Intervention for [Client Company]. Retrieved June 21, 2008, from http://www.workplacecoaching.com/pdf/Coaching.ComReport.pdf

Wadsworth, A. (2001). Analysis of the use of Internet-based communication technology for online post-training coaching. ProQuest Digital Dissertations, (UMI NO. AAT 3017240), retrieved January 23, 2004 from http://80-wwlib.umi.com.libproxy.sdsu.edu/dissertations/fullcit/3017240

Walther, J. B. (1993). Impression development in computer-mediated interaction. *Western Journal of Communication, 57,* 381–398.

Walther, J. B. (1995). Relational aspects of computer-mediated communication: Experimental observations over time. *Organization Science, 6*(2), 186–203.

Walther, J. B. (1996). Computer-mediated communication: Impersonal, interpersonal, and hyperpersonal interaction. *Communication Research, 23*(1), 3–43.

Walther, J. B, & Burgoon, J. K. (1992). Relational communication in computer-mediated interaction. *Human Communication Research, 19,* 50–88.

Wang, L. & Wentling, T. (2001, March). *The relationship between distance coaching and transfer of training.* Paper presented at The Academy of Human Resource Development, Tulsa, Oklahoma. Paper retrieved June 21, 008, from http://learning.ncsa.uiuc.edu/papers/coach.pdf

Wasylyshyn, K. M. (2003). Executive coaching: An outcome study. *Consulting Psychology Journal: Practice and Research, 55*(2), 94–106.

Whyman, W., Santana, L., & Allen, L. (2005). Online follow-up: Using technology to enhance learning. *Leadership In Action, 25*(4), 14–17.

Wilkins, B. (2000). *A grounded theory on personal coaching.* Unpublished doctoral dissertation, University of Montana, Missoula.

Zaccaro, S. J., & Klimoski, R. J. (2001). The nature of organizational leadership: An introduction. In S. J. Zaccaro & R. J. Klimoski (Eds.), *The nature of organizational leadership: Understanding the performance imperatives confronting today's leaders.* San Fancisco, CA: Jossey-Bass.

SECTION II

ORGANIZATIONAL CASE STUDIES

CHAPTER 9

E-MENTORING

Opportunity to Build a Virtual
Learning Community with Teachers

Leena Vainio and Irja Leppisaari

ABSTRACT

As virtual work and education become more common, support for personnel
development should incorporate virtual methods. The mentoring process can
be expanded by means of new digital tools to form virtual mentoring communities. In the Finnish Online University of Applied Sciences, e-mentoring was
tested during a project where the aim was to evaluate and develop the pedagogical quality of learning content and virtual teaching and learning. Training
on evaluation criteria application was arranged in diverse mentoring groups
at the beginning of the project. The study modules were evaluated alone, in
pairs, or in groups, by using the criteria of pedagogical quality evaluation. The
mentor was involved in the evaluation discussions as much as possible. Each
group decided upon the evaluation methods to be used together with the
mentor. The most crucial outcome was the pedagogical discussion supported
by e-mentors about the quality of and collegiate support for online education.
Shared evaluation criteria were a good tool for focusing the conversation on
the actual theme and pedagogical quality was approached from various view-

Virtual Coach, Virtual Mentor, pages 177–187
Copyright © 2010 by Information Age Publishing
177

points. The project demonstrated that e-mentoring can be used to support the development of teachers' online pedagogical competence through national virtual peer networking. E-mentoring that is integrated into the content of a development project offers a fresh and effective virtual learning space to teachers. The combination/integration of expert and peer mentoring seems to be an essential element of the Virtual Mentoring Community.

INTRODUCTION

Collective expertise seems to produce the best solutions to the increasingly demanding challenges posed by working life (Bereiter & Scardamalia, 1993; Hakkarainen, Palonen, Paavola, 2004; Nonaka & Takeuchi, 1995). Currently, communities bolstering expertise emerge particularly frequently in the social media of the Internet environments. Even though the majority of social media communities are established in connection with spare time and hobbies, occupational communities are emerging, allowing experts to solve problems together, exchange ideas and often also to develop innovations.

In the past few years, a teacher's tasks have increasingly shifted from solo performance to networking. Technology has added a new dimension to networking in the information society. However, technology is not the most essential feature in this society but rather the new way of working creatively based on human interaction. Virtual work, e-work, distributed work, networked organization, and distributed teams are today's reality in many companies and organizations (Vartiainen et al., 2007).

As virtual work and education become more common, support for personnel development should incorporate virtual methods. The peer network is considered a significant learning resource, which has brought the development of group and peer mentoring into the spotlight (see Clutterbuck, 2004; Klasen & Clutterbuck, 2004). As an interaction process peer learning is more dynamic and equalizing than traditional mentoring (Colky & Young, 2006). The mentoring process can be expanded by means of new digital tools to form virtual mentoring communities. In this case we are illustrating e-mentoring as a method of developing the competence of teaching staff and collegiate work. E-mentoring is in this context defined as a process of developing and sharing expertise, mainly in interaction taking place in the virtual environment.

AIMING AT A LEARNING COMMUNITY

According to Lewis and Allan (2005), features of learning communities are a shared goal and project; commitment to the improvement of professional practice; learning and development focused on real work-based issues and practice; high levels of interaction and collaboration, knowledge sharing, construction and exchange; and the use of information and communication technologies.

In this case we are describing how e-mentoring was used to support teachers' online-pedagogical competence in the peer network in the Finnish Online University of Applied Sciences. The networked way of working and sharing and development of expertise between organizations mounts the central challenges to this joint union. In our case the activities can be labelled as the Virtual Mentoring Community (VMC), since the operations were managed, members represented diverse organizations, and knowledge and skills were transmitted between the Virtual Mentoring Community and the teacher's workplace. The operations were also characterized by a collaborative approach to problem solving as well as a communal and boundary-crossing way of working. In addition, the project had a common development focus (specific work-related task): pedagogical quality evaluation of the study modules within the Finnish Online University of Applied Sciences. The participants were offered the necessary virtual communication tools, and by means of them they could flexibly communicate either synchronously or asynchronously. At the same time, the ways of working and the environment provided support for the development and evaluation needs of individual University of Applied Sciences in virtual education (see Lewis & Allan, 2005; Leppisaari, Mahlamäki-Kultanen, & Vainio, 2008). The project formed a *learning space* that came into existence within the community as the individuals communicated and exchanged ideas, knowledge, experiences, and emotions in a reflective and authentic way (see Docherty, Boud, & Cressey, 2006), which leans on collegiality as a learning resource.

BACKGROUND OF THE PROJECT

Four hundred and fifty teachers in 50 production teams produced learning objects and online courses collaboratively at the Finnish Online University of Applied Sciences during 2004–2006. The Finnish Online University of Applied Sciences is a networked co-operation organization formed by all 29 Finnish Universities of Applied Sciences. The outcome of the project was approximately 500 credits of online courses and approximately 400 learning objects by the end of 2006. The material is developed in production teams made up of teachers of Universities of Applied Sciences arranged by educational area and specific subjects.

During the project one aim has been to develop the pedagogical quality of the learning content and virtual teaching and learning. The best way to increase the quality of virtual learning is to develop the virtual teaching competences of the teachers. The Online Pedagogy and Research project during 2005–2007 has supported the production teams in the creation of pedagogically high-quality learning objects and study modules. The tools deployed in the development are the criteria of pedagogical quality evaluation and e-mentoring of study modules.

PROJECT OBJECTIVES, ALLOCATION OF RESOURCES, AND PARTICIPANT ROLES

In this case we depict the project completed in 2007 that set out to assess the quality of study modules, bringing forth learning communities in collaboration with teachers by relying on e-mentoring. The objectives of the venture were:

- to make study modules created at the Finnish Online University of Applied Sciences known and promote the use of materials in individual University of Applied Science
- to develop pedagogical quality evaluation criteria for study modules and test them in practice
- to design a personnel development model based on e-mentoring

As assessors, the project recruited teachers from Universities of Applied Sciences for each of eight learning branches. The purpose was to involve those teachers who had not actively been engaged in producing material. The interfaces of the Finnish Online University of Applied Sciences managed recruitment from within their own University of Applied Science, involving one teacher per organisation. Twelve Universities of Applied Sciences (of a total of 29) undertook the venture, and altogether 34 teachers commenced, while three discontinued the evaluation during the project. Fifty man hours were allocated to the teachers for their contribution, but the project did not compensate for the evaluation work. The Universities of Applied Sciences enabled most of the participants to carry out their project tasks during their working hours, since the venture translated into continuing education. The majority of the teachers took up project tasks entirely on a voluntary basis and were not allotted any special resources. Some of the teachers were "obliged" to participate in the project, which clearly had a negative impact on commitment and the learning and collaboration process.

The teachers were compressed into six evaluation teams. Experts in online pedagogy were invited as group mentors, their task being to act as discussion partners in issues related to quality evaluation mainly in interaction taking place on the web. The tools used in interaction and the evaluation sessions were the Moodle learning platform for asynchronous interactions and the Adobe Acrobat Connect Professional (Connect Pro) communication system for synchronous interaction, both of which offer the possibility to share and write together documents and use video picture and sound.

Traditionally mentoring has been regarded as an honor, and the motivation to take on mentoring has been based on the benefits acquired by the mentor through expansion of one's own expertise and networking. The project did not compensate the mentors for their efforts. They operated voluntarily or acquired funding from their own organizations to support mentoring, the scope of which was set at 40 man hours.

Two managers were appointed to this e-mentoring project. One of them was, in particular, responsible for the objects of the operations and the implementation of the pedagogical quality evaluation of the study modules, and the other for the personnel development side of the project. Both were adept at the development of online pedagogy and e-mentoring related to University of Applied Sciences, with previous experience from project collaboration. Both were allotted 50 man hours per month to focus on the project.

THE PROCESS OF E-MENTORING AND TRAINING

The mentoring project was implemented during a ten-month period (January to October, 2007). At the beginning of the year, two months were spent on testing the tools, developing the evaluation criteria, and recruiting mentors and participants. The actual evaluation and mentoring took place during March through September, and the closing seminar and final reporting in October; the process was interrupted by the two-month summer vacation period.

The evaluation was performed mainly online as peer evaluation in groups headed by the mentors. The evaluation groups had the option of arranging a face-to-face meeting with their mentors, and the financing would have been provided by the project, but only one of the groups took up this opportunity.

Mentors and teachers were able to test the operating environments before the actual get-togethers, and during the first virtual meetings, technical support was available full-time. A dedicated support person was appointed for both software tools, to be contacted whenever problems arose. The Moodle learning environment served as a common "memory space" where all the general issues such as instructions and pedagogical quality criteria were stored. In Moodle each mentor had his or her own working space, where the mentors could build communal working environments such as discussion forums or wikis. Furthermore, each group had an online virtual meeting room in Connect Pro where they could conduct their team meetings. Most of the team meetings were recorded (Connect Pro software has this possibility) and the recordings were stored in the Moodle environment.

Training on the application of the evaluation criteria was arranged in the diverse mentoring groups at the beginning of the project. The study modules were evaluated either alone, in pair, or in groups, by using the criteria of pedagogical quality evaluation. Each teacher evaluated the material first alone, followed by pair or group discussions, and the final evaluation was based on these common evaluation outcomes. The mentor was involved in the evaluation discussions as much as possible. Each group decided upon the evaluation methods to be used together with the mentor. In addition, in their own preliminary evaluation the teachers assessed the study modules they had selected in their own institutions together with their colleagues. A comprehensive view of the study modules under scrutiny was thereby acquired. At the same time, study module contents were made known on a wider scale in various Universities of Applied Sciences. The most crucial outcome was, however, the pedagogical discussion supported by e-mentors about the quality of and collegiate support for online education. Shared evaluation criteria were a good tool for focusing the conversation on the actual theme, and pedagogical quality was approached from various viewpoints. In the final virtual meeting in September 2007, each evaluation group drew a summary of its activities and outcomes as well as listed further development recommendations for the material. The results were presented in the project's closing seminar, also arranged entirely virtually through Connect Pro. This allowed flexibility for the busy teachers in terms of time and location.

The project consisted of the following sub-areas:

- familiarization of project managers with virtual operating environments in small and bigger group situations
- guidance in the use of digital tools and virtual environment (all the participants had the opportunity to test the operability of the environment before the get-togethers)
- orientation for mentors through Connect Pro
- orientation for teachers through Connect Pro (also the mentors took part)
  - general session introducing project objectives, ways of working, evaluation criteria of pedagogical quality, and groups and e-mentors
  - finally each mentor took his or her group to their own virtual meeting room and the group planned the way forward
- meetings chaired by mentors (approximately 3–5 online meetings and the additional discussions in the Moodle environment)
- messages and e-mails through Moodle, face-to-face meetings (once per each group)
- guidance and consultation by project managers for mentors
  - joint virtual meetings for mentors and project managers once a month

  – each mentor had the opportunity to invite project managers to the
    meetings for support (attendance in all the groups at least once)
- surveys supporting self-evaluation conducted at project start and
  end, targeted to both teachers and mentors
- closing/evaluation seminar in Connect Pro, with both teachers and
  mentors participating

EVALUATION OF THE PROJECT

Research and development have been integral elements of the project all
throughout the process. The initial questionnaire surveyed the actors' starting points and expectations. The monthly discussions between the mentors and project managers focused on gathering information based on self-
evaluation, and the process was improved accordingly. A particularly large
portion of feedback concentrated on the evaluation criteria of pedagogical
quality, both contents-wise and in relation to their implementation as an excel sheet. The criterion set was adjusted in accordance with the feedback.

The mentoring methods were continually developed on the basis of the
feedback. Interesting issues were raised in discussions during the process—
in particular, concerning the impact of virtual environment on dialogue.
The mentors found e-mentoring extremely challenging. It was also easy to
evaluate one's own activities, as the mentoring sessions were recorded (videos by Connect Pro), and afterwards one's performance could be observed.
The mentor's roles received special attention: Is he or she a coach, networker, discussion partner/discussion activator, or developer of e-mentoring?
One of the mentors concluded he or she was more of a discussion activator
and that when it comes to the other roles, his/her activities as a mentor
required yet more practice.

The teachers were mainly satisfied with the mentoring in the project,
and they felt that benefiting from it was largely dependent on their own
efforts. Eighty-three per cent of the teachers who responded to the final
survey considered the mentoring to have offered support in the evaluation
process. As one of the teachers expressed: "The mentoring surprised me
positively. The mentor's expertise helped me crystallize my thinking, but
he also listened to our messages with respect for our insights. Encouraging expertise." It was the mentor's task to boost the process with questions
supporting reflection: "...by asking questions and also taking part in the
conversation he activated us to discuss with each other."

Teachers had problems with finding sufficient time to commit to the interactive project and evaluation work. Especially in those cases where their

own institution had not allocated actual resources for the work, people sometimes discontinued the effort.

COMMUNICATION

As communication tools the project deployed principally virtual tools: the Connect Pro communication system and the Moodle learning environment. The feedback from virtual communication was surprisingly positive. The majority of participants did not know each other beforehand, but the virtual environment, which offered the possibility to benefit from video picture and sound, allowed the participants to get to know each other.

The virtual environment also made the attendees stick to the point and focus their attention on what was essential. The teachers found the tools chosen for the project appropriate for e-work within the network. More practice was called for to gain fluency in web-based dialogue: "More should be invested in personal dialogue instead of so-called careful, overly sophisticated and fact-based conversation, which at times made e-working troublesome. It would be necessary to promote dialogue more by means of versatile tools." Working in a virtual environment will necessitate more practice. The team leader should remember to moderate the discussion evenly to allow all the participants and attendees to actively voice their insights. Listening to the other party is crucial also in the virtual environment.

OUTCOMES AND LESSONS LEARNT

The aim was to achieve collegiate group mentoring. The goal was completely met in only three groups out of six, as the teachers found it difficult to find time together to interact. Also, the project groups' work was critically interrupted by the summer vacations. Timing, scheduling, and specified work methods have clear impact on development ventures of this type. The time allocated to this project was too limited as time was needed also for the study of methodology and for getting familiar with the evaluation criteria of pedagogical quality and with study module contents.

The groups conducting either pair or group evaluations reached a more in-depth level in their pedagogical discussions. When evaluating material characteristics, also one's own e-teaching skills and related development areas were analyzed; plenty was learned from one another. The mentor's role and ability to create an interactive learning space impacted the conversational build-up. In groups where evaluations were performed alone, the mentor remained largely no more than a provider of instructions winding up the discussion.

However, it can be observed that the virtual way of working in itself does not induce the above-mentioned problems; on the contrary, it enhanced the opportunities for working together and exchanging ideas. The participants regarded the method as an enabler of flexible work, but scheduling was the major hindrance. At times also, technical problems inhibited interaction: Computer configurations did not work in the virtual meetings, and sometimes no sound or picture was available. Attendees in projects of this type should have access to technical support also within their own work community, as not all problems can be solved remotely.

Another problem was the fact that not all of the mentors had previous experience with mentoring, and with e-mentoring, in particular. The role of the mentor in this project was to lead the discussion towards the development of online pedagogical expertise. Now some of the mentors remained on too "instrumental" a level, getting stuck on evaluation criteria or other content-related details, and philosophising on online pedagogical issues, so that the development of teachers' own online pedagogical competences had to assume a minor role. One solution could be pairing mentors up, with one more experienced and another less experienced partner balancing each other. The mentor pair could build more multifaceted dialogue among the participants. Sharing the responsibility would also unleash more energy for raising the contextual substance once the resources are not tied solely to organizing group activities and meeting schedules.

Project outcomes:

1. Common quality evaluation criteria were defined to be benefited from in the future and as early as in the content production phase.
2. Plenty of information was acquired about the quality of online material created at the Finnish Online University of Applied Sciences and their development needs.
3. Experiences were gained regarding the functioning of virtual environments in e-mentoring.
4. The participants learned from one another, the reciprocal development and sharing of expertise among colleagues was intensified, and one's own online pedagogical competence was developed.
5. Development aspects were formed for the implementation of e-mentoring (peer and groups).
6. Virtual ways of working became part of teaching routines.

The following conclusions can be drawn on the basis of the venture outcomes (cf. also Leppisaari, Mahlamäki-Kultanen & Vainio, 2008):

- E-mentoring that is integrated into the contents of the development project offers a fresh and effective virtual learning space to teachers.

The combination/integration of expert and peer mentoring seems an essential element in its creation.

- The mentor has four essential tasks/roles in the build-up of a common learning space VMC: *sparring partner, networker, discussion partner, and e-mentoring developer.*
- Expert and peer mentoring supported the opportunity to learn in action together with experts and colleagues and simultaneously develop the activities under scrutiny.
- Teachers in particular expect clear operating models in a virtual environment. Mentoring is expected to support communal working, by means of which learning can be intensified in the project. Therefore the integration of the project's contextual learning objectives into operating methods enhancing the quality of interaction in the virtual environment will emerge as the key development challenge in the VMC.
- In the future, concrete ways of working promoting dialogue in e-mentoring should be modelled, developed, and rehearsed in orientation training directed to e-mentors (cf. Clutterbuck, 2007).

All in all, the teachers reported that the project integrated contextual and online learning in a motivating way, educating them in both the evaluation of pedagogical quality and the utilization of new tools and methods. The project demonstrated that e-mentoring can be used to support the development of teachers' online pedagogical competence through national virtual peer networking. Further studies will help develop operating models promoting the creation of a social process when striving for more in-depth dialogue and sharing of expertise both with the mentor and the colleagues in virtual mentoring groups.

REFERENCES

Bereiter, C., & Scardamalia, M. (1993). *Surpassing ourselves: An inquiry into the nature and implications of expertise.* Chicago: IL: Open Court.

Clutterbuck, D. (2004). *Everyone needs a mentor* (4th ed.). London: Charted Institute of Personnel and Development.

Clutterbuck, D. (2007). Making mentoring work in an international environment. In I. Leppisaari, R. Kleimola, & E. Johnson (Eds.), *Kolme säiettä kasvuun: verkkopedagogiikka, koulutusteknologia ja työelämäyhteys* (pp. 246–260). Kokkola: Central Ostrobothnia University of Applied Sciences.

Colky, D. L., & Young, W. H. (2006). Mentoring in the virtual organization: Keys to build successful schools and businesses. *Mentoring & Tutoring 14(4)*, 433–447.

Docherty, P., Boud, D., & Cressey, P.(2006). Lessons and issues for practice and development. In D. Boud, P, Cressey & P. Docherty. (Eds.), *Productive reflection art work* (pp. 193–206). London and New York: Routledge.

Hakkarainen, K., Palonen, T., Paavola, S., & Lehtinen, E. (2004). *Communities of networked expertise.* Earli. Amsterdam: Elsevier.

Klasen, N,. & Clutterbuck, D. (2004). *Implementing mentoring schemes.* Oxford: Elsevier.

Leppisaari, I., Mahlamäki-Kultanen, S., & Vainio, L. (2008). Virtuaalinen ryhmämentorointi ammattikorkeakouluopettajan kehittymisen tukena (Virtual group mentoring supporting a teacher's professional development). *Aikuiskasvatus (Adult Education) 28*(4), 278–287.

Nonaka, I. & Takeuchi, H. (1995). *The knowledge-creating company. How Japanese companies create the dynamics of innovation.* New York: Oxford University Press.

Vartiainen, M., Hakonen, M., Koivisto, S., Mannonen, P., Nieminen, M. P., Ruohomäki, V., & Vartola, A. (2007). *Distributed and mobile work: Places, people and technology.* Tampere: Otatieto.

CHAPTER 10

THE BRIGHTSIDE TRUST

Dr. Tessa Stone

SUMMARY: OVERVIEW OF THE BRIGHTSIDE TRUST

The Brightside Trust (www.thebrightsidetrust.org) is a educational charity which provides online resources and tools to help individuals make informed choices and overcome barriers to education and employment. The charity focuses on providing e-mentoring support to individuals who need information, skills and confidence to overcome social disadvantage. Brightside has been running successful e-mentoring projects for over five years in the UK, working with a range of cross-sector partners, including government, companies, charities and universities. Since 2003, over 12,000 people have participated in e-mentoring with Brightside, approximately 7,500 mentees and 4,500 mentors.

E-mentoring builds on the widely recognized benefits of face-to-face mentoring, such as raising motivation and increasing confidence. Electronic mentoring ("e-mentoring") enables mentees and mentors to communicate via an online platform, which is thus independent of geographic or scheduling constraints. This offers participants flexibility and is also a way to add value to face-to-face activities by extending the support and maintaining momentum with participants.

Virtual Coach, Virtual Mentor, pages 189–198
Copyright © 2010 by Information Age Publishing
189

Brightside's e-mentoring offers two types of support: online information, and access to an e-mentor to help mentees navigate and use this information and to provide encouragement.

> The feedback and advice from my e-mentor has really boosted my confidence. She has made me realise that I do have all of the characteristics required to be a successful applicant for medicine....I feel that without her help and advice, I would not feel this ready for applications in October. (Mentee on Bright Journals scheme)

BRIGHTSIDE'S APPROACH—INNOVATION IN E-MENTORING

Traditionally, e-mentoring has been conducted via email. Brightside's e-mentoring platform uses "blogging," which enables mentoring pairs to communicate through personal online journals. Brightside has also pioneered the use of "content-driven" e-mentoring which enables users to integrate their conversations with information resources and interactive coaching activities on the site. This enables e-mentoring interactions to be stimulated and structured by online content, which has been tailored to their individual needs and to the objectives of the e-mentoring project.

The e-mentoring platform offers three components:

- Conversation
- Content
- Coaching

Conversations

Once participants have created an online profile and been matched to their mentor/mentee, they can communicate via their online journals. This secure conversation space ensures mentors and mentees can have confidential discussions. Once participants have been posted a message, they will receive an email alert. Mentors can have up to four mentees, and can hold individual conversations with them all, or start a group conversation. Mentors can also communicate with other mentors on their scheme via a group conversation.

Content

Each of Brightside's e-mentoring websites has an online library containing resources and topical news articles. This content is written by Brightside's web editing team to disseminate relevant and engaging information to the user, provide a stimulus for conversations between mentors and mentees and to maintain engagement in the project. The content is written specifically for the user group and on average there are 60 new articles each month. For example, for projects aiming to help young people access university, articles about student finance, the university application process and exam revision will be uploaded at the key points throughout the academic year. Users can comment and feedback on articles and there is also an 'Ask an Expert' function which allows users to submit queries on a specific topic to be answered. RSS news feeds pull content from other relevant websites and the library serves as a portal to other information sources.

Coaching

The third area of the website provides a curriculum of online coaching activities, which enables mentors to assign relevant tasks for their mentee at regular intervals. Mentors can review and comment on their mentee's progress on a task; thus providing structure to the mentoring and acting as conversation prompts. These activities are developed to help mentees develop key skills that have been identified or learn about specific themes. For example, young people on Brightside e-mentoring projects can work on building a budget, or writing their CV with their mentor.

Brightside believes that there is more to e-mentoring than the provision of a website. In order to achieve the engagement of mentees and mentors, and therefore the most positive outcomes, project coordination is vital. Brightside works in collaboration with its partners to achieve this. In addition to the delivery of the e-mentoring platform, Brightside provides:

- web content to keep the websites fresh, engaging and relevant
- training sessions and materials for mentors and mentees
- ongoing monitoring, moderation and technical support
- project management and implementation advice

SECURITY

Each mentor and mentee is provided with a secure login ID and a password. Conversations and passwords are encrypted between browsers and servers. All communications are filtered to identify offensive or abusive language, e-mail addresses, and mobile phone numbers. Any transgressions are "paused" in the system, and cannot be published until cleared by Brightside monitoring staff. In addition to the automated checking processes, communications are randomly assessed and read by human eye (the Brightside staff and/or other project coordinators). If participants feel they receive an inappropriate message on the site, they also have the opportunity to report it individually.

All mentors must complete security checks including fully enhanced disclosure (CRB), before they are granted authorization to use the website.

BRIGHTSIDE PROJECTS

The Brightside Trust is currently running over 30 projects, with a particular emphasis on supporting young people into and through higher education. There are several programme strands:

- **Progression to higher education**

  Young people face confusing progression routes, financial challenges and stiff competition for university places and jobs. Brightside works in partnership with universities to deliver e-mentoring projects which link 14–18 year olds with an e-mentor who can inform them, inspire them, and help them prepare for the transition to university and beyond.

  These projects enable young people to be mentored by an undergraduate who can offer 'near to peer' advice. Mentors can share their own experience and discuss key issues about higher education with them, such as course choice, study skills, university life and student finance. Some universities develop a customized project with Brightside that is bespoke to them, while other institutions subscribe to one of Brightside's existing projects to enable their students to take part.

- **Subject specific e-mentoring**

  Brightside has experience in running e-mentoring projects aimed at encouraging young people to consider careers in specific professions, including engineering, law, architecture, medicine and the sciences. Mentees are supported by mentors who are studying the subject at university or by mentors from the particular profession. The websites provide targeted news and information about these subjects as well as career profiles, interviews and specific resources.

Science is often perceived as being impenetrable, the work of The Brightside Trust is a great example of how young people can be excited and inspired to enter these fields.

—Professor Lord Robert Winston

- **Enterprise challenge through e-mentoring**

  The "Big Deal Blogs" project uses e-mentoring in a specific, time-limited context. The project involves school students in a 10-week enterprise challenge where they compete in teams to develop a business idea. Each team is supported by an e-mentor from business who helps them develop their idea, write a business plan and pitch the idea to the judging panel in a final presentation.

- **Retention and employability at university**

  In partnership with a university, these projects aim to support undergraduates to complete their course and make a successful transition to the world of work. New undergraduates are mentored by final year students to support them through the transition to independent living and university life. Final year students can be mentored by mentors in industry who can help them prepare for graduation and job application processes, looking at interview techniques, the job search and CV writing.

- **Transitional e-mentoring for targeted groups**

  Brightside also provides e-mentoring support for targeted groups, including children in care, and ex- and young-offenders. Mentors in these projects are trained to support mentees in the following areas
    – Accessing employment or training
    – Developing their employability skills
    – Building their confidence and self-esteem
    – Make a successful transition into independent living

Conclusion

The Brightside Trust has over five years of experience in developing and implementing e-mentoring projects. We firmly believe that this model of intervention has the potential to achieve real change for individuals. At a time when competition for university places and jobs is fiercer than ever and progression routes for young people are many, varied and complex, there is a need for flexible, personalised support. For those people who don't have access to the information they need, and cannot rely on their own social or family networks, e-mentoring can offer effective, valuable, inspiring support. Brightside aims to continue to innovate in technology and to work collaboratively with partners to ensure that e-mentoring support is expanded to help those who need it.

E-MENTORING RELATIONSHIP CASE STUDIES

Case Study 1

Jordan (mentee) and Claire (mentor) were paired up in November 2007. They were participating in a project which aims to support young people from socially disadvantaged backgrounds to enter the competitive fields of medicine, dentistry, veterinary science and the allied health professions. Their relationship from the very beginning was a practical one. Jordan was already confident and focused in what she wanted to do—she had planned work experience, had made the right A-level choices for Medicine, and was certain that she wanted to go to university.

> **Jordan:** I was thinking about doing child nursing, but when I started at [my school] and told my tutor that, he said (and careers people have said) that I was setting myself too low, because of my grades I could achieve better. I don't think I want to be a paediatrician though.

However, Jordan had been given conflicting advice by careers advisors at her school, and had little or no information about what she needed to do to get onto a medicine course specifically.

> **Jordan:** At the moment I am looking at university prospectuses and information . . . what other places did you apply for and what did you consider while choosing them? There are also lots of open days upcoming, so I shall go to some.

Claire was able to offer her first-hand advice on the type of course she should take to cater to her interests. Also, most importantly, Claire walked her through all of the necessary components for a medical course application including work experience, different UCAS rules for medicine, UK-CAT examination information, and bursaries for the UKCAT.

> **Claire:** Have you had any advice regarding the UKCAT? I think you will have to get ur name down to sit it before too long. I am currently in the process of writing a website (with some other people from my course) in order to give advice to people applying to med school. A guy who's in first year and has sat the paper prepared some information on it, which I have copied and pasted at the end of my post as I thought it might be of use to you.

This is the kind of invaluable information that most students would get from their parents; however, as Jordan was the first person in her family to go to university, no one in her family could help her to prepare. Claire was able to talk her through the process, and Jordan points out that it is good to have a record of her advice online so that she can look back over it and review, and ensure that she doesn't forget anything useful.

Other mentees from this project stated that:

> My mentor has been really helpful. Not only has she discussed with me the details of med school years so far, she has also helped me develop as a person. (Bright Journals mentee, 2007)

> I felt that it was very encouraging to know that there was always someone there that I could talk to about the difficult times at college and the advice I was given was very helpful. (Bright Journals mentee, 2007)

Case Study 2

The following case study illustrates the importance of tone and boundaries when mentors post messages.

Mentor Yvonne works as a chemical engineer and has two mentees, Jodi who in year nine and was preparing for her SATs and Evelyn, who was in year 11 and in the process of applying to sixth form college.

Many mentors and mentees adopt an informal, friendly style of writing, sometimes swapping chatty information about their lives and day-to-day worries. Yvonne tells her mentee Evelyn about her busy week, and Evelyn responds much as a friend or colleague would.

> **Yvonne:** Things have been so busy at work and at home recently that I'm all over the place. We've awarded a whole bunch of contracts at work that has taken ages, I've had a plumbing disaster in my flat (a leak from my bathroom meant the neighbour's bathroom ceiling fell down!) and my partner has changed jobs—all at once. Still, at least I won't get bored which is, as you know, one of my pet hates!
>
> **Evelyn:** Aww, sorry to hear about it. You must be absolutely knakered. Pus you've your own job to deal with! Good luck on everything!

Although many mentors are quite a bit older than their teenaged mentees, communicating through a website appears to allow participants to ignore the age gap in a way that might not always be possible face to face. While initial posts by mentees tend to make references to age, later messages show mentees and mentors communicating as equals and chatting about their interests and sharing their experiences. The potential for informal chat appeared to create a relaxed atmosphere in which mentees felt comfortable asking questions and exploring new ideas.

> **Jodi:** Wow, Japan in 24 hours! Sounds . . . crazy!! Did you sleep? We did a project on Japan in geography. It was so interesting because it's so physically and culturally diverse. It's a range from massive sky scrapers to beautiful mountains and lakes!! It was a really interesting project and I'd love to see the place, though I'm not a seafood lover! But they don't all eat sushi, I guess!

Mentor training is critical to ensure mentors understand the remit of their role as mentor. For example, the mentor training highlights the importance of avoiding telling their mentees what to do but instead making suggestions about things that their mentee could take into account when making important decisions. This encourages and empowers the mentee to take responsibility for their own choices. After being asked by her mentee which 6th form college to choose, Yvonne aims to give her mentee, Evelyn the guidance to make her own decision.

> **Yvonne:** I guess the questions you need to ask yourself are whether the college is the right one for you in terms of the other subjects you want to study and the 'vibe' you feel when you visit it. Do you think you will be happy there? Do you get a good feeling about the teaching staff and the other students? It is important that it's convenient and the facilities are important too—if it's a good place to be you'll feel more motivated. Does any of that help? I'm a great believer in trusting your instinct.

In addition, Yvonne admits that however good the advice given, mentees don't always have to accept it.

> **Yvonne:** Having said that, I have a confession that I actually only did Chemistry and Physics at O level (what is now called GCSE). Of course, it was all different then—I'm not quite a dinosaur but it was quite a long time ago!

In this way she gives Evelyn the space to listen to advice but ultimately make her own decision.

Both of Yvonne's mentees gained confidence through this relationship. Yvonne developed a good relationship with both of her mentees through discussion that was not only limited to education topics, but also a range of issues that brought an informal tone to their conversations. For mentees, this provided an informal atmosphere in which they can feel confident to raise any questions they may have. It also means that the mentor can give her advice as an equal rather than an authority figure—the mentees seem more than happy to take it on board.

Case Study 3

John (mentor) and Ruth (mentee) were matched in November 2007 on a Brightside project which was set up to increase participation in physics at university by increasing the subject-awareness and raising the aspirations of non-traditional university entrants,

Ruth initially viewed the mentoring relationship as an opportunity to discuss the specifics of physics with John.

> **John:** Have you heard of CERN? They study particles which collide at ridiculous speeds and energies and can recreate conditions that were around only thousandths of a second after the universe evolved!

As the relationship developed it became clear that, while enthusiastic about physics and further study, Ruth was unsure about university itself. She was worried about the workload, course options, and living arrangements, and she had many misconceptions about university life.

> **Ruth:** Do you ever have so much work you get extremely frustrated? and I've been wondering what happens if you don't get an assignment in on time? and do you have detentions?

With examples from his own experience, John talked Ruth through a typical week at his university, and explained the different options in terms of course and module choices. After this, the conversation changed direction—the assumption now was that Ruth would definitely be going to university, and she began to attend open days.

> **Ruth:** We got shown around an Oxford University college. We then did a tour of Oxford City Centre to see some other university colleges. This was mainly for the benefit of those that were here from outside Oxford. We also were given the chance to eat amongst other current university students.

The relationship then moved into a different phase where Ruth posted less frequently, but with more focused questions about concrete issues such as course choice or university choice.

Note: The names and some details in these case studies have been changed for data protection reasons.

CHAPTER 11

CONNECT ASSIST COACHING CASE STUDY

Rusty Livstock

This case study describes a nine month pilot programme to offer coaching support to teachers in training and newly qualified teachers across Wales. The opportunity arose due to a Welsh European Social Fund programme that Connect Assist was part of. Connect Assist is the three year old multi-channel contact centre that provides all service responses to the various teacher support charities.

In 1999 the English Teachers Benevolent Fund created The Teacher Support telephone line to offer personal and professional support to teachers. A similar service was extended to Teachers in Wales in 2002.

Teaching is a profession which is deceptively pressurised. It is estimated that one in three of those that train never go on to teach, while of those, that enter the profession two out of every five will last less than 5 years.

Teacher training is intensive with direct classroom observation being a vital assessed element. The early years as a teacher are regulated by an evaluated induction programme and a series of assessments alongside holding down an initial job. Teachers are often difficult to attract into initiatives so it was felt that the combination of a supportive programme and the individual support offered by coaching would be an initiative that should be

Virtual Coach, Virtual Mentor, pages 199–208
Copyright © 2010 by Information Age Publishing
199

investigated as an aid to retention. We set a target of enrolling 4% of the total population of 4000.

It was felt that it would be interesting to investigate the viability of coaching in impacting on the retention of teachers, so we constructed an offering for teachers in training and newly qualified teachers.

Connect Assist is an occupational coaching provider that creates affordable coaching for organisations through use of a web based tool. This coaching tool is located behind customer organisation websites, via an application program interface.

The coaching tool allows the coachee to interact with a named coach in a password protected web area. There is space for recording goals, journaling and storing articles of relevance. The coach has a management consol that uses a traffic light system to monitor frequency of interaction with coachees. This allows multiple clients to be managed at the glance.

To initiate this approach we needed the support of the Teacher Support Cymru (TSC) Charity. Together with them we worked out a promotion strategy that saw announcements on websites and presentations in all training colleges. Finally there was an e-mail announcement, explaining the programme to the target audience. The response from both TSC and the trainee teacher population was very positive.

We agreed a process for coachees and designed a sign on assessment survey and an additional fact sheet. The coaches sent out invitations to those that had signed up at training college workshops.

The offering was:

- Free development coaching
- A central point for getting advice and support for the conflicts that confront them.
- The focus could be personal or professional.

We anticipated offering 6 hours per individual after an initial phone taster session; participants were free to use the service via the internet or phone.

Our Coaches are full time employees who work around a 24 hour, seven day week shift pattern. They have extensive education understanding and to complement them we recruited a Welsh speaker with career guidance experience.

Training for the team consisted of training from officers of the Teacher Support Charity. Various team members attended the recruitment days in colleges and from this dialogue we built a first hand expectation of what the group might want. This proved especially helpful to the contract coach, for whom it offset her relative lack of education sector experience.

Our most significant issue at this stage was one of semantics as new and trainee teachers said that they were not interested in coaching as, in the context of their training, coaching meant dealing with problems in curriculum learning. This proved to be a somewhat intractable issue as the obvious alternatives, mentor or coach, have other specific meanings in teacher training.

CONNECT ASSIST COACHING CASE STUDY (V3)

The time from project agreement to implementation was unrealistically short. Planning occurred in January and in February we researched potential user thoughts and opened enrolment. When by late March no one had formally enrolled with us we generated an email flyer to all who had been seen at presentation meetings and following this people began signing up

Coaches reported that candidates were:

- Wanting to chat.
- Often using the phone rather than just on line contact
- Particularly appreciating direction to information downloads
- Seeking assistance across a breadth of topics, almost too broad to categorise.
- Frequently appearing reluctant to define goals and get engaged in actions focused on progress.
- Only very infrequently requesting services in Welsh.

In our initial planning we had envisaged that by late May we would develop specific project related materials. To date this has not materialised as demand has never justified it.

In May and June a stress workshop was run at training centres and at support sessions. These sessions were received overwhelmingly positively and had the unforeseen consequence of significantly boosting the Cmyru Coaching membership.

RESPONSES/TAKE UP

Initial take up of this scheme was as the extract from our CRM records shows 830—a phenomenal improvement on pre-project expectations.

Incidents

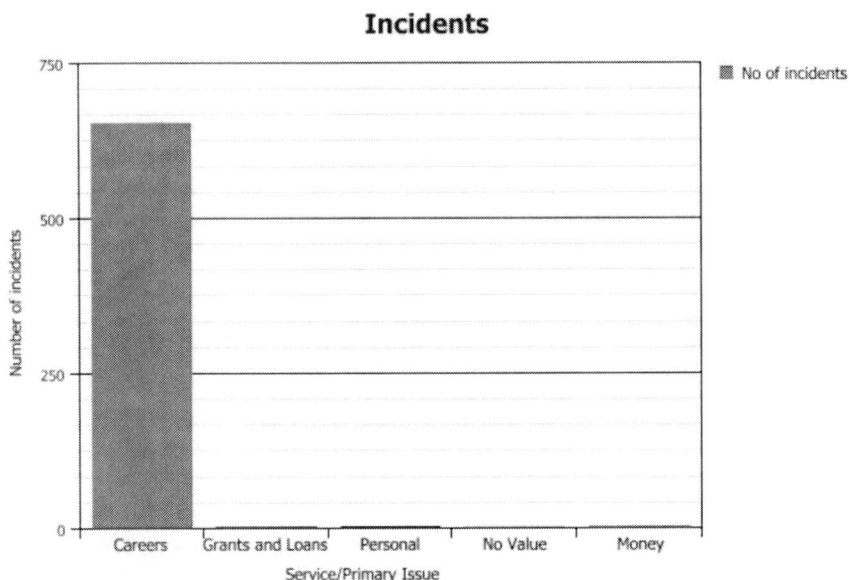

Figure 11.1

But regular usage of the service dropped to a disappointing 385. This is atypical of other services that we run and so merited investigation.

Particularly poor use seems to have occurred in the group who signed up following a promotional talk/and email.

Individuals who signed up following all of the above and a stress workshop session are subjectively perceived to have been better engaged and motivated. It's difficult to judge accurately on this but it appears that education regarding management of self for purposes of improved performance helped the potential coachee to better understand what was on offer and therefore to connect with it.

This project has produced information that we believe will help us do things differently in any future schemes.

SERVICE USE IS TO DATE AS FOLLOWS

Incidents by Primary Issue

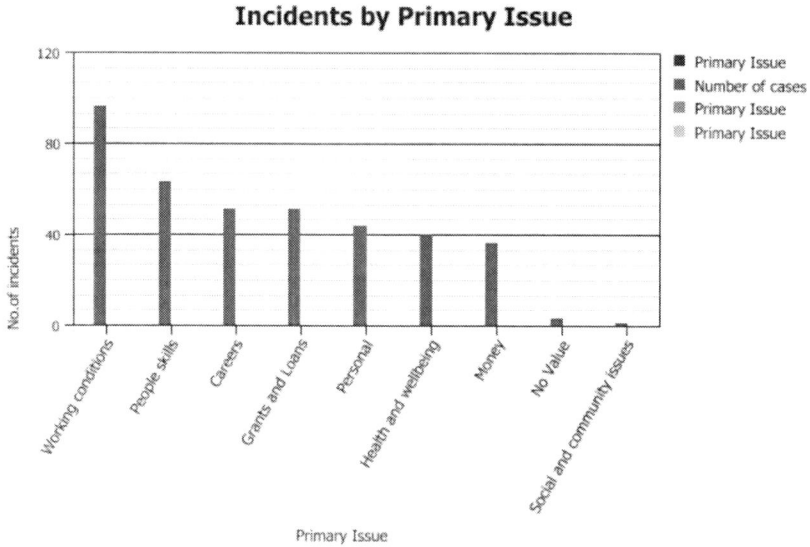

Figure 11.2

We found that in the top issue of *working conditions* concerns were more specifically (Figure 11.3):

Incidents by Primary Issue

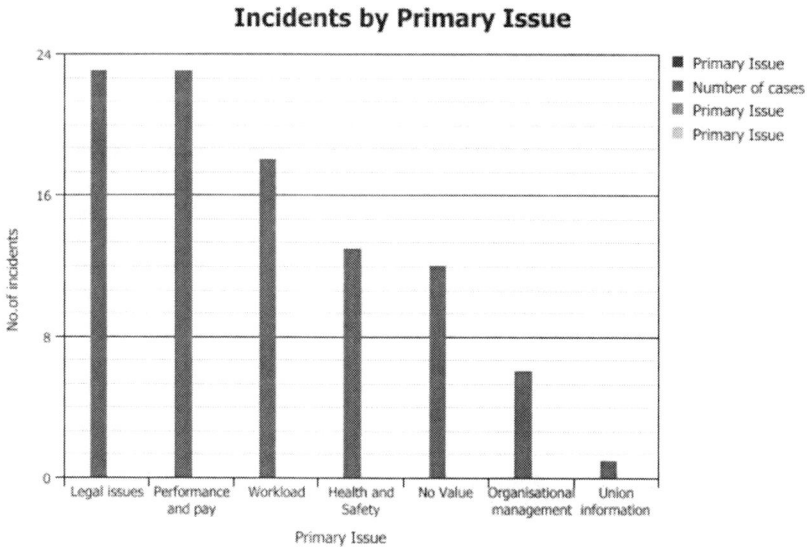

Figure 11.3

Below those working conditions focused items, relationship issues were the coaches' primary focus (Figure 11.4).

Incidents by Primary Issue

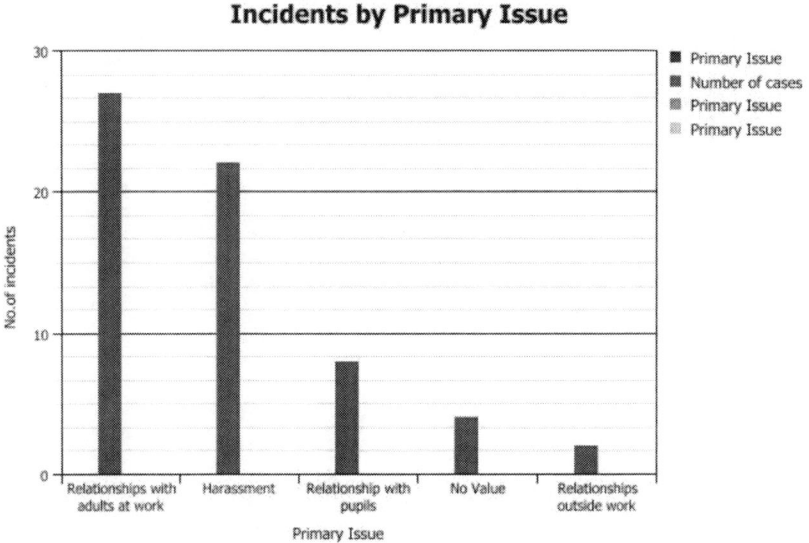

Figure 11.4

This data is consistent with that for teachers across the entire range of charities so it did not surprise us.

TWO ANONOMISED COACHING VIGNETTES

Example 1:

Thank you for joining the SOC Online Coaching Service.

Answer: There is no time limit following the award of Qualified Teacher Status (QTS), within which NQTs have to start their induction. However, it is of course always best whenever possible to start induction soon after being awarded QTS so as to build and strengthen the skills gained during initial teacher training.

With regard to trying to finding a suitable placement to do your induction you can try contacting the LEA's in the areas within which you would like to work. Some LEA's, but not all, work in partnership with local supply agencies and can help you to find the right position. The Teacher Training

Agency will be able to provide you with further information and you can contact their LEA Induction Co-ordinator as follows:

I hope this information is helpful for you and wish you luck in finding a suitable position.

Best Wishes.

Example 2:

Thank you for completing your survey.

Can I assure you that while we work together any communications between us are confidential, (there are some rare exceptions—for example where someone is in danger of significant harm to themselves or others)

Have you been able to talk to anyone at work about the difficulties you are experiencing?

I note that you are concerned about your performance and finding it difficult to manage etc.—has anyone else raised any concerns?

It would be really helpful if you could let me know what, if any, support you have sought to date, and what the results of this have been, and how you feel I can assist you from a coaching perspective.

In the mean time, I have assigned some articles to you which may be of interest.

Best Wishes.

Many thanks for your swift reply!

With regards to the issues you have mentioned in your email, I have spoken to my manager about my concerns and she has said that I do not have anything to worry about and that she is happy with me. However, I always doubt what I do and I find myself second guessing what I do. This means that I end up procrastinating about my work and then getting stressed about lesson planning. It is also impacting on my teacher training as I am now behind with my lesson plans for that. I know that I probably sound like a bit of a whinger but I have now got to a point where everything to me seems wrong. I know that teaching is a big responsibility but I am finding that a bit difficult to manage at the moment. I enjoy working with students but I am now realising how much work is involved and I am not sure that I can deal with it.

206 ■ R. LIVSTOCK

I would like a way of being able to deal with my negative thinking and perhaps some organisational strategies?

I have also been to the doctor and I have been given medication to help with my sleeping and I have a follow up appointment to decide on how to treat my anxiety.

I look forward to hearing from you soon!

It's nice to hear from you again.

From what you have said I think you might find the following web site interesting:

www.mindtools.com

It covers just about most things including stress/anxiety/building self esteem/positive thinking etc.

Have a look and let me know what you think—perhaps you could try one or two of their strategies and see how you get on.

I can procrastinate with the best of them—I am also very good at doing the easiest of tasks first! What I found worked for me was to make a time table for myself—start off with one thing that I didn't mind too much—then one of things that I would normally put off again and again and then reward myself with something I liked to do! It's important to try and break things down into small achievable portions so that you do not become overwhelmed by the size of the task in front of you.

There are lots of teaching resources available via the internet. The Teachernet website is really useful and the following websites may be of interest regarding lesson plans:

www.lessonplans4teachers.com   www.lessonplanspage.com

OBSTACLES/LESSONS LEARNED

Our initial hypothesis, that coaching could be used to reduce workforce attrition, will take 2–3 years to test. In the meantime the project has produced considerable insights which are transferable across the entirety of the education sector and across all public sector employment, if not all professions.

Everyone here is aware that services with a personal development focus are always best received when pulled in response to an articulated demand. This "Starting out Cymru" scheme seems to further bear out this well known fact.

Our efforts at researching what our population might perceive as helpful have to date proved inconclusive and we have acted on all requests without any distinctive success.

We were pleased that the service is used with an appropriate focus on workplace effectiveness. It remains the case, that it is a negative experience that more often seems to drive users to us. But as generations of dentists would tell us, the public rarely are sufficiently preventative in their tooth brushing why should approaches to mental health be any different from those to dental health!

We are clearly doing some thing's correctly with a 20+% sign up rate looking like great insurance provision returns and 8.5% utilisation as being highly respectable for a sign yourself in, self service scheme.

Our coaches (as you can see from the online vignettes supplied) contract, seek clarity, empathise, and offer resources.

Our Coaches would argue that people in this cohort could have had benefited from greater

- focus
- follow through

Reactions to the service were positive but not ecstatic.

The perceptual issue of coaching as "coaching is what you get if you have a problem" is perhaps uniquely difficult with young UK based teachers. Coaching can be a difficult concept to convey theoretically, when the offering is internet based that communication is even more problematic.

We do use our internet and phone modalities highly successfully with other clients these young pressurised professionals seemed to be greatly assisted into the coaching by an element of face to face and trialling it opportunities.

Time is the resource that no modern professional sees themselves as having enough of, our teachers were no exception and the unevenness of demand in their working year appeared to exacerbate their patchy use of the resource.

A high number of the email addresses supplied by students in the cohort, were of the variety that is perceived by system firewalls as spam. This wasted staff time, caused frustration and lost us some coaches.

Our technology when accessed from a website takes two clicks and then involves the completion of a survey. Our initial findings suggest that two is a click too many and that a survey that has more than five questions significantly deters usage.

Going forward we would use our e-marketing facilities more regularly to send products to engage users. With a young audience such as incoming professionals, it would also be beneficial to have the development funds and time to add supporting podcasts to the offering.

We will continue to work with this group until the end of the year.

We would like the opportunity to work with similar professional groups on the issue of: "what can be done to assist retention in professions especially of expensive to train essential public employees?

We will in particular, seek to study what leads or disinclines users to enrol in and stay engaged with an online coaching scheme.

CHAPTER 12

ASKMAX.CO.UK
WEB-BASED MENTORING

Dr. Angus I. McLeod

AskMax is a mentoring product using the World Wide Web as a vehicle for providing stimulating challenges, thoughts, and ideas to cohorts of staff. Each 'Max' is a selected coach, fully versed in the methods and signed up to our best-practice charter. Each coach will typically be assigned six to fifteen mentees. Max is 'invisible' to the mentee, so cover is easily arranged if our coach has to be away for a time. All contacts by mentees have to be responded to within twenty-four hours in the working week.

The first organization approached was in 1999 and it is in the high-volume retail sector. The organization was undergoing a great deal of change in their supply chain, and AskMax was seen as a valuable way of achieving rapid healing of issues amongst staff throughout the UK.

GETTING THE BEST FROM A MENTORING APPROACH

Where ideas, stories, and advice are given, three options are provided if possible. Because brains are adapted to rapid comparing, two solutions would typically lead to a preference for one solution provided by the mentor.

Virtual Coach, Virtual Mentor, pages 209–213

When the mentee is given three solutions, thinking becomes more complicated as the comparing process requires several comparisons sequentially:

1. A + B
2. B + C and possibly,
3. A + C

As the comparing begins, the mentee will typically switch from comparing and start to process at a more creative level. This helps mentees to get away from yes/no reactions and to think more productively, sometimes finding their own solutions or an adaptation of one of the solutions put forward by the mentor. We expect greater motivation to arise from the mentee's solutions than from any of the mentor's suggestions.

Max is genderless, and when a mentor is unavailable, their list of mentees is fluidly taken over by another Max (e.g., Max AA., Max AB, etc.).

UNIQUE SELLING PARAMETERS

The program was sold to the L&D Manager and the Director of Supply Chain HR. It was understood that staff who were emotionally involved with issues would have the benefit of having to be logical about the issues in order to be able to describe their situation to Max—itself an immediate, potential benefit in both wellbeing and productivity. As so much damage to the working community was coming from management itself, the provision of a wholly third-party service was seen as a bonus. Almost no support was required by the organization, and summary reports giving information about the types of issues coming forward would allow management to implement support structures and to plan learning and development initiatives for the future.

IMPLEMENTATION METHODOLOGY

A pilot group of executives were offered the opportunity to take up the AskMax product. They were given a basic summary of the objectives of the offering, how the system would work and over what time-frame. For those wanting more information, example situations were provided showing how the email-based mentoring method can help executives through a range of issues. Forty-four executives were placed on the pilot list and about half of them (21) used the service within the first three month period. The company budgeted for total usage (44) but was only required to pay for the number using the service in each three-month period.

DATA

AskMax provides a summary of issues by category, and we worked with the organization to adapt and agree on suitable headings for these categories. Each mentor would be categorizing each individual issue within these headings to provide data for the summary. From the first three month period, 67 individual requests came to the five AskMax mentors. All mentors were experienced coaches and all familiar with the requirements of the service.

The average response time of mentors was 28 hours 11 minutes (including weekends and twenty-four hour working).[1] The average mentor work-time was just under 27 minutes with the shortest and longest work-times being 4 and 65 minutes, respectively. Mentors were paid on an agreed, fixed basis (per mentee using the service, within each three-month period).

The categories of individual issues (19 of them) were grouped as shown in Table 12.1 under six headings. These groups are shown together with a percentage for the incidence of that group of category arising.

The majority (68%) of users only used the service for one category of issue within a three month period with some bringing two issues (22%) and just two people bringing more than two issues (three and four respectively).

The number of mentor replies needed for any given category of issue to reach an agreed completion are shown in Table 12.2.

TABLE 12.1

Group	Percentage
Relationships	24
Role/Job	44
Performance Review Issues	4
Training	1
General Development	15
Other	12

TABLE 12.2

Mentor replies needed "x" to complete	Number of incidents needing "x" replies
1	14
2	8
3	8
4	2
5	1

ANECDOTAL FEEDBACK FROM MENTEES

"Speaking to somebody that I either don't know or doesn't know me is great."

"Thank you for your useful service."

"This opportunity has allowed me to re-focus on my current role and the weaknesses that I have identified and to relate the role I am now in to some of my successes in my previous roles."

"Meant that I actually did something about the issue, I was avoiding it."

"It allowed me an external view of the issues and helped me put a plan of action together to cope with the situation. Though the issue is still not resolved I feel 100% better about it and feel that it will be resolved in the future."

"The great benefits of the process are in its simplicity and impartiality (no hidden agendas) i.e., simply having someone that is not related to the company asking you questions, challenging your thinking and providing advice and guidance."

"Have used the e-mail service and have found that it excellent."

"The responses have been of an extremely useful nature."

"The process which you suggested I go through before the meeting was tough but my manager was so impressed with it that he is looking to spread it throughout the team as part of our development process. Thank you, Max."

"The advice you gave has been invaluable."

"Thanks again for some great advice, I shall put all of your recent e-mails into a plan of action."

"Thanks for your reply, it is definitely one of the most useful e-mails I have ever been sent."

CONCLUSION

The AskMax model works very well and in this pilot met expectations for both the business and a majority of users. The only changes made since this early pilot have been client-specific. The uptake of the service was higher than expected (21 from 44) reflecting, possibly, the general stress-level in the workforce at the time. The uptake may well have been even greater had staff been permitted to self-select into the service rather than being nominated as a group. The third-party confidentiality of the service appeared to provide a trustworthy prospect for almost half those offered the service even though they were selected by management. The AskMax mentoring model should not be as challenging as face-to-face coaching but in any case provides (for many people) the opportunity to detach emotionally from an issue and to logically present it to a third party. Once mentoring began, research and actions by staff began to be taken, sometimes after a long period of inaction and distress over an issue.

NOTE

1. Later studies included calculations with the weekend break subtracted. Typical response times on that basis then reduce substantially, with one cohort in another organisation enjoying average response times of 4 hours, 18 minutes.

CHAPTER 13

E-MENTORING @ STAFFORDSHIRE UNIVERSITY

Janet Wright and Jean Simpson

INTRODUCTION

The e-mentoring scheme was introduced to the Faculty of Sciences at Staffordshire University for the 2006–07 academic year, and it involved four subject areas: Biological Sciences, Forensic Science, Geography, and Psychology and Mental Health.

The e-mentoring scheme is part of the wider University strategy to improve retention and as such, its purpose is to provide new undergraduate students at Staffordshire University with access to an extra layer of support and guidance at minimal cost to the University. It is a student-centered peer-mentoring scheme that uses electronic means of communication to link first-year students with more experienced students (the e-mentors) for one academic year. These e-mentors can provide new students with practical help in adapting to University life and enable them to gain a better understanding of the university culture and increase their awareness of the facilities that are available. As such it provides a vehicle to deliver additional support, tailored to the requirements of individual students, even within subject areas with large student numbers (see Hixenbaugh, Dewart, Drees,

Virtual Coach, Virtual Mentor, pages 215–222
Copyright © 2010 by Information Age Publishing
All rights of reproduction in any form reserved.

& Williams, 2005, for a greater discussion), while enhancing both social and academic integration into the university community.

There are two key outcomes to the e-mentoring scheme at Staffordshire University:

1. To offer all first-year and direct entrant students within the subject areas involved in the scheme with an effective peer support system through which they are provided with advice and guidance from someone who understands what it is like to be a new undergraduate student at the University.
2. To provide second- and third-year students within the subject areas involved in the scheme with training and experience of mentoring, which is an invaluable life skill that they can use to both help others and to enhance their own CV.

Over the past decade the growth in electronic means of communication has had a considerable influence on personal interactions and the development of social networks. Many new university students are already familiar with social networking websites such as Facebook and consequently feel comfortable using a variety of electronic communication media. Further, electronic communications are often perceived as providing a safe, non-intimidating, and flexible communication space in which students can discuss subjects that they are not always comfortable talking about in a face-to-face situation (Dewart, Drees, Hixenbaugh, & Williams, 2003; Hamiliton & Scandura, 2003; Hixenburgh et al., 2005). Thus combining peer-mentoring with electronic communication systems provides the opportunity to develop peer-support networks in a format that is relevant and familiar to the twenty-first century student. The template and guidelines developed as part of an HEFCE-funded Widening Participation project by the Department of Psychology at the University of Westminster (Dewart, Hixenbaugh, Thorn, & Drees, 2004) were chosen as the format for the e-mentoring scheme at Staffordshire University.

HOW DOES THE SCHEME WORK?

The scheme involves providing all first-year students with the opportunity to develop an e-mail partnership with a second- or third-year student from their subject area, who becomes their e-mentor for the year. A web site has been set up to provide additional support for, and information about, the scheme (see www.staffs.ac.uk/schools/sciences/ementoring). A flyer introducing the e-mentoring scheme and directing the reader to the web site for further information is included in the Welcome pack sent out to applicants after they have confirmed their acceptance of a place at the University.

The web site largely follows the template provided by the University of Westminster (see Dewart et al., 2004, and www.wmin.ac.uk/sshl/page-444). It provides a general overview of the scheme, information about the scheme specifically for the mentors and mentees, details of the e-mentoring team, a link to the e-mentoring bulletin board, a summary of the University IT policy relating to use of emails and electronic discussion boards, details of and links to useful resources (both links to University support services and external sites about mentoring), and the e-mentoring agreement used by the scheme.

It is a voluntary scheme; first-year students can opt out of the scheme if they do not wish to participate in it, and all the e-mentors are volunteers drawn from the subject areas participating in the scheme. All mentors are provided with training on how to be an effective mentor and in how the scheme works before they take up their mentoring role. Both the mentors and mentees are required to sign an e-mentoring agreement. This agreement is a modified version of the one devised by the University of Westminster (Dewart et al., 2004) and has primarily been designed to protect the mentors and the mentees. It provides definitions of an e-mentor/mentee and covers issues of confidentiality and inappropriate communications (see www.staffs.ac.uk/schools/sciences/ementoring/ementoringagreement/ for a copy of this agreement).

The main way in which mentors and mentees communicate with each other is via their University email accounts, and e-mentors are expected to respond within two working days to any e-mail queries sent by their mentees. Mentees are paired with mentors during the University Welcome Week, and mentors are required to send a welcome message to their mentees by the end of the first teaching week. Each e-mentor is responsible for mentoring between six and twelve first-year students.

As well as communicating via email, a bulletin board has also been set up to provide an alternative means of communication and extra support (see www.staffs.ac.uk/schools/sciences/ementoring/bulletinboard). All members of the e-mentoring scheme (e-mentees, e-mentors, and e-mentoring staff development team), together with the tutors, support staff from the subject areas involved, and staff representatives from Student Support, are able to both access and post contributions on the bulletin board. This board can be used to post queries about any aspect of life at Staffordshire University (academic and social), announcements relating to events at the University (academic and non-academic), and information relating to specific subject areas and individual modules. The board is organized into several different forums: a general forum that all users can access, and a forum for each subject area involved in the scheme. Tutors and support staff can access all the forums, but mentees and mentors can only access the general forum and the forum relating to their specific subject area.

WHO IS INVOLVED IN THE MENTORING SCHEME?

Three main groups are involved in the e-mentoring scheme: the staff development team, the student e-mentors, and the e-mentees.

The staff development team consists of two project leaders (an academic and an administrative officer), tutor representatives from each subject area, the faculty IT manager, and a representative from Employability and Student Support at Staffordshire University (see www.staffs.ac.uk/schools/sciences/ementoring/theteam/). At various points during the development and implementation of the scheme, additional support was provided by appropriate staff from across the University. Tutors from each subject area, particularly first-year personal tutors and module tutors, are encouraged to take part in the scheme via the bulletin board.

RECRUITING AND TRAINING E-MENTORS

All the e-mentors involved in the scheme are volunteers and no re-imbursements or other incentives are offered to them, apart from the opportunity to enhance their CV. Any second- or third-year student who volunteers and undertakes the required training is eligible to become a mentor. All volunteers who complete the academic year as a mentor will receive a certificate that they can include in their personal development portfolios.

Emails are sent to all first- and second-year students within the participating subject areas mid-way through teaching block 2 inviting them to volunteer as e-mentors for the next academic year. Students are asked to register their interest in being involved in the scheme by replying to this email. These students are then invited to a subject-specific mentor recruitment session at which they can discuss the role of e-mentors with appropriate staff. Following this, on confirmation of their intention to participate in the scheme as e-mentors, students are invited to attend the training session held in September.

During the first year of the e-mentoring scheme mentor training was delivered via two events: a compulsory full-day event at the very start of the academic year, and a shorter optional session mid-way through teaching block 1. Volunteers were required to attend the main training day in order to be able to participate as e-mentors. This event included sessions on "Theory and Practice of Mentoring," the facilities and services offered by Employability and Student Support, the services offered by the Students Union, guidelines for e-communication and netiquette for the e-mentoring scheme, how to use the e-mentoring bulletin board, and how the scheme works within specific subject areas. The second optional training event followed up the "Theory and Practice of Mentoring" session and included "Resolving Issues within Mentoring Practice," "Managing Expectations as a Mentor," "Mentoring Health Checklist," "Tips on E-mentoring Emails," "Introducing the GROW Model," and "Ending Mentoring Relationships."

EVALUATING THE E-MENTORING SCHEME

During 2006–07 the effectiveness of the e-mentoring scheme was evaluated in the following ways:

Mentee Evaluation:

1. To assess how effectively the scheme had been introduced to the first-year students and to check that welcome messages had been received from e-mentors, nine questions relating specifically to the e-mentoring scheme were added to the Welcome Week Evaluation that all first-year students are required to complete early in teaching block 1.
2. First-year students from each subject area were asked to complete a questionnaire evaluating the scheme at the end of both teaching blocks.

Mentor Evaluation:

1. A questionnaire was distributed via the bulletin board and email in mid-October which asked the mentors to provide some feedback on (a) the effectiveness of the training session, and (b) how they felt the scheme was going.
2. At the end of the optional e-mentor training session in late November, a focus group session was held in which mentors' views on how the scheme was going and how it was organized were obtained and mentors had the opportunity to raise any issues or concerns that they had about the scheme.
3. At the end of teaching block 2 an end-of-year questionnaire was distributed to all mentors via email.

The Mentees' Experience in 2006–07

The evaluations of the mentees' experience indicated that the scheme was generally well received by the first-years. Although only a relatively small number of the respondents indicated that they had used the scheme regularly or personally benefited from it (16%), the majority of the respondents (>85%) were very supportive of the e-mentoring initiative and indicated that they thought the scheme was a good idea and should be made available to future first-year students. Students using the scheme were happy with the timing of mentee-to-mentor allocation and the number of messages that they received from their mentor, but would like these messages to be more carefully targeted at certain points in the year, such as just before coursework submission deadlines or class-tests. The most popular reasons for contacting their e-mentors were to ask for advice about meeting coursework deadlines, some clarification about the structure or organization of their awards, and advice on finding off-campus rented accommodation for year 2. Generally responses by their mentors to these queries were prompt and helpful. The personal benefits of the e-mentoring scheme identified by mentees included providing reassurance about the whole induction process and relieving both general anxiety about being at University and more specific anxieties related to coursework and examinations.

The Mentors' Experience in 2006–07

Evaluation of the scheme by the mentors was extremely positive. They all felt that it was a very valuable initiative and should continue to be offered to future first-year students. Most of the respondents (>95%) felt that they had received adequate and appropriate training for their role as an e-mentor and were provided with sufficient support during their time as a mentor from the faculty. The majority of them indicated that they responded to their mentees' queries within two days, and most of them (>80%) reported that they felt able to offer appropriate support and guidance to their mentees throughout the year. Mentoring responsibilities did not interfere with their studies; generally mentors spent up to one hour per week on their mentoring duties with none spending more than two hours. Benefits of mentoring included improving communication skills, learning how to listen to others and give advice, satisfaction of helping others, and increasing their self-confidence. The mentors' suggestions for improvements to the scheme included "Consider delivering the main training component earlier and in shorter sessions," "Mentees should be allocated to mentors earlier (before welcome week)," and "More contact between mentors should be encouraged."

The Use of the Bulletin Board

Both evaluations by the mentors and mentees, together with general monitoring of the use of the bulletin board, showed that this was the least successful element of the e-mentoring scheme. During the first year of the scheme neither the mentors nor the mentees made much use of this facility. Mentees indicated that they preferred to contact their mentors directly for advice, while key information about modules and notification of subject area, faculty, and university events were routinely distributed via email and module VLEs such as Blackboard.

AREAS FOR CHANGE

Following evaluation of the first year of operation of e-mentoring within the Faculty of Sciences, the main alterations to the scheme for the 2007–08 academic year were a revision of the mentor training provision, the use of more carefully targeted emails, and an attempt to increase e-contact between mentors. There is still a compulsory training event at the start of the academic year, but this is now shorter; the participants are sent information with which to familiarize themselves prior to the event and are provided with greater opportunities to discuss "What if…?" scenarios during the training. After their welcome message, mentors were asked to try to relate the timing of their subsequent messages to periods when first-year students within their subject area are likely to be most anxious or stressed (e.g., submission of their first pieces of summatively assessed coursework, class-tests, first oral presentation, etc.). In order to both facilitate increased contact between mentors and increase the use of the bulletin board, a dedicated mentors' discussion forum was set up on the e-mentoring bulletin board. Although mentor evaluations suggested that mentor-mentee relationships should be initiated prior to the mentees formally starting at the University by both holding the main training session and allocating mentees to their mentors earlier, this was felt to be impractical as not all the mentors would be available for training and to take up their mentoring duties prior to Welcome Week. Formal evaluations have yet to be carried out for the second year of the scheme; however, observations by staff involved in implementing it indicate that mentors would benefit from further advice from first-year tutors on appropriate timings for targeted emails to their mentees. Also, use of the bulletin board is still low, suggesting that the continued inclusion of this facility in the scheme may need to be reconsidered.

CONCLUSION

Although the Faculty of Sciences did show improved retention rates for the 2006–07 academic year, it is not possible to identify how much of this improvement can be directly attributed to the introduction of the e-mentoring scheme, as this initiative was one of several measures, such as a strengthened personal tutoring system that was introduced to improve retention rates within the faculty. Evaluation of the first year of operation of the e-mentoring scheme has shown that it has enhanced the learning experience of many students within the faculty, both mentees and mentors. Mentees valued it as a mechanism for reducing stress and anxiety associated with the various facets of university life, while mentors recognized the important role that it played in increasing their self-esteem and self-confidence. As it is an extremely flexible and accessible but low-cost means of providing additional support to a large number of students during their most vulnerable period at university, e-mentoring will continue to be an integral part of the retention strategy within the Faculty of Sciences at Staffordshire University for some time.

REFERENCES

Dewart, H., Drees, D., Hixenbaugh, P., & Williams, D. (2003). Electronic peer mentoring: A scheme to enhance support and guidance and the student learning experience. *Proceedings of the Educause in Australasia 2003 Conference, 734–737*. Retrieved March 14, 2007 from http://www.caudit.edu.au/educauseaustralasia/2003/EDUCAUSE/PDF/AUTHOR/ED030049.PDF

Dewart, H., Hixenbaugh, P., Thorn, L., & Drees, D. (2004). *E-mentoring: Information Pack*. London: University of Westminster.

Hamilton, B.A. & Scandura, T.A. (2003). E-mentoring: Implications for organizational learning and development in a wired world. *Organizational Dynamics, 31(4)*, 388–402.

Hixenbaugh, P., Dewart, H., Drees, D., & Williams, D. (2005). Peer e-mentoring: Enhancement of the first year experience. *Psychology Learning and Teaching, 5*, 8–14.

Staffordshire University (2006) E-Mentoring: Faculty of Sciences. Retrieved September 30, 2006 from http://www.staffs.ac.uk/schools/sciences/ementoring/

University of Westminster (un-dated) Department of Psychology: e-mentoring. Retrieved September 30, 2006 from http://www.wmin.ac.uk/sshl/page-444

SECTION III

INDIVIDUAL CASE STUDIES

CHAPTER 14

INDIVIDUAL CASE STUDIES

INTRODUCTION

This chapter is intended to balance the perspectives we have taken on virtual coaching and mentoring. In the first part of the book, we looked at the research evidence and the generic practicalities of making these distant relationships deliver value as a means of learning from others. In the second section, we explored some organisational case studies, illustrating some of the breadth of approaches and applications of this flexible and adaptable medium.

Here, in this final section, we take the view of the participant in e-mentoring or e-coaching relationships. The style of presentation of each of these cases differs considerably, and this reflects the diversity of the individuals and their relationships.

RAHIB—E-MENTORING

How and why did this relationship come about?

I was introduced to Zulfi through a mutual acquaintance and kept in moderate contact for some time. I called on Zulfi in 2005 when we launched the Inclusion & Diversity initiative for our global company. Zulfi gave an inspirational speech about his life experiences and how he had developed a reputation for being "the crazy one"—someone who never gave up and impossible didn't exist in his dictionary. Many of the 120+ attendees found

Virtual Coach, Virtual Mentor, pages 225–240
225

Zulfi's speech inspirational, especially as it resonated with their experiences as ethnic minorities living in the UK.

Many things about Zulfi reflected my own status:

1. Ethnicity (Pakistani British)—both born in Pakistan and grew up in Yorkshire
2. Religion (Islam)
3. Education (graduated from British universities)
4. Corporate success (both had progressed well in previously nationalized utilities, although Zulfi has been more successful)
5. Interest in inclusion/diversity and mentoring

From this positive experience, we developed our relationship further.

What was the initial contract? Did it change over time? If so, how and why?

Initially, the relationship was informal and without any form of contract as we often called one another discussing a whole spectrum of things—ranging from personal aspirations to reflecting on the state of the (professional, corporate, political) world.

Very soon—within a couple of months—it soon dawned on me that this relationship could be much more beneficial and that I should better utilze my time with Zulfi. Reflecting on our relationship, identifying Zulfi as a confidante and recognizing his extensive skills in mentoring and coaching, I proposed to Zulfi that I would like him to mentor me on a semi-formal basis. After some discussion, we developed some basic rules:

1. I would drive the agenda.
2. I would make contact each time.
3. To start, I would make contact on a fortnightly basis.
4. We would plan our discussions before making contact.
5. In emergencies, I could contact Zulfi via phone.

What mix of media did you use?

At the start of the mentoring relationship, most discussions were via telephone, usually in the late evenings (outside working hours and kids in bed), lasting 30 minutes to an hour each time. This helped to build a strong foundation and rapport and deepened understanding of one another's value to the relationship.

From time to time we communicated via MSN when I felt that I would benefit from writing things down instead of rambling on the phone.

As time progressed, we discussed meeting—however, due to demanding professional and personal schedules, a number of planned meetings had to be cancelled. Being a bit of a traditionalist, I believe face-to-face meetings are important. As I was participating in a video conference at work, it occurred to me that Zulfi and I could talk via video call on MSN Live Messenger. So when I got a web cam as a Christmas gift, I was very pleased!

We quite often e-mail one another during the day.

Text messaging is also useful when travelling by train and when I need a short response to a spontaneous question.

Which media worked best/least?

Contact by mobile telephone works well when I am on the move—except when travelling by train because we often get cut off and flow of conversation is not good. A long battery life is required when discussions go on and on and on! The ears get hot also so a hands free device works better.

Text messaging is extremely useful for short, swift advice—but not for lengthy discussions.

E-mail works best when I feel I need to carefully articulate my issue/concern and when I feel I need a considered response from Zulfi—this is particularly important when a major decision regarding lifestyle or career needs to be made. In these cases it has been important for me to carefully develop my thinking to enable Zulfi to help me arrive at a pragmatic, balanced solution/decision. It also gives Zulfi some thinking time. However, e-mail is not the best solution if you have an urgent issue that you feel needs immediate attention because the addressee may not attend to the email for some time. E-mails can be used to maintain a record of conversations that can be referred to or reflected on from time to time.

Video call by MSN Messenger Live works best when I want to engage Zulfi from the comfort of my home without any distractions—face-to-face contact is important and perhaps more inspiring as you can see expression behind the words used. And it's free to use!

How did the virtual environment compare with your experience of face-to-face developmental alliances?

I think most people would agree that face-to-face relationships work best. However, I have been pleased with the MSN video call, because as long as there are no distractions, it is like having a face-to-face conversation. With technological developments apace, there are more powerful products becoming available on the marketplace. These products continue to improve audio and video quality, and with broadband, the transfer of data is fast enough to enhance the engagement experience.

I have found e-mentoring to be extremely powerful, especially when we have challenges with distance and time.

A tool that records the conversation and then provides a transcript would be ideal—maybe for the future!

Can you give us a flavor of the discussions?

During the mentoring relationship, we have been able to navigate through a number of critical junctions in my life. I would highlight one as an example: I reached a point in my career where despite my best efforts, it was clear to me that I would have to make a major decision to make a career move. The company went through a reorganization that resulted in a reduction in overall numbers, and I did not fit in the future plans of new management. I had the choice of struggling through or taking control of the situation and managing my career. With Zulfi's help and direction and a methodical, consultative approach, I was able to:

- Understand what was important for me—personally and professionally
- Understand what my aspirations were
- Understand levels of control I could exercise (full, part, little, none)
- Realize where I needed help and how it could be secured
- Assess risks/rewards

With this approach, I was able to:

- Understand my journey
- Consider and analyze my options
- Make confident decisions through analysis of my research
- Focus on the best option
- Plan each step to achieve
- Secure a great move into an area of the business that reflected my aspirations and had a fantastic fit with my skills and experiences.

What concerns, if any, did you have about the relationship?
How were these resolved?

I was conscious that this was a stressful time for me and it would have been very easy to become reliant on Zulfi. I did not let this happen. I made sure that I did not contact him more than once a week and for a maximum of 30 minutes per call. I spoke with him and he helped me to arrive at my own decisions and actions during these engagements. At times where I could not make progress, we would agree for me to do further analysis, or think a little more, or take time to decide between difficult options. Through questions, Zulfi helped me to clarify my thinking and to distill options bearing in mind what was important for me.

Zulfi was very understanding and generous with his time. I had to be disciplined not to take advantage of his generosity. I always knew I would end up in a better place and wanted to grow my relationship with Zulfi and not destroy it.

This approach continued for a couple of months.

What were the outcomes? How, if at all, did they differ from the
original goals?

In the end I secured a position at the same company, but in a different and new part, unrelated to my previous position. Looking back, I often wonder how I ended up in this part of the business. It takes me back to what Zulfi said to me some time ago: "Never forget a friend!" The position I secured came about when I congratulated an old friend on a successful promotion—I took him for a coffee to talk about his new role, the challenge, and how I would be keen to join him. A week later, I became a senior member of his team.

What would you do differently?

Reflecting on this tricky period, I would change a number of things:

1. Take action quicker—opportunities pass by as quickly as does time.
2. Engage my mentor early—this helps with a methodical approach and without the emotion.
3. Take control of the situation—do not rely on or expect others to do it for you.
4. Focus on the output—o secure a better position in the quickest time.

What lessons did you learn?

Through this experience, I have learned many lessons:

1. It is important to control one's emotions and make decisions on sound analysis of options.
2. Do not take things personally—many people face career challenges through business transformation.
3. Engage trusted advisers—two heads are better than one—and those close to you know you better than you know yourself sometimes.
4. Remain flexible and open-minded—opportunities can spring up from the most unexpected of sources.
5. Trust your instincts—and strike while the iron is hot.
6. Use technology as an aid to reach others—ride a horse because it's quicker and easier than walking!

KATHLEEN FROUD—E-COACHING

How and why did this relationship come about?

The client was moving from a full-time position with an organization he had been with for over 20 years to setting up his own business as a Business Improvement Coach and was looking for help in channeling his thoughts, setting goals, and taking action to achieve them. He was aware that he some-times needed someone to challenge what he was doing and that this should be someone who was independent of his clients, family, and friends.

What was the initial contract? Did it change over time?
If so, how and why?

The original contact and contracting were undertaken by e-mail, and is-sues were clarified and confirmed in the first telephone session.

The first e-mail to the client introduced the coach and gave some back-ground on her experience in coaching/mentoring and asked the following questions:

- What would you like to achieve from the on-line sessions, and what are your expectations?
- How frequently would you like to make contact, and what is the most convenient time during the week for this to occur?
- How would you like to make contact: telephone, e-mail, or a mixture?
- If you would find it useful to have a telephone chat before we engage in an on-line dialogue, please let me have your telephone number and details of when it is convenient to contact you.

The client's response to this provided a good background to the client's expectations, confirmed the regularity of contact and preferred media plus current e-mail and telephone contact details. Preferred day and time of contact each week were also indicated.

The contracting was revisited at various times through the process of coaching as challenges and difficulties arose. For example:

- What should happen when either party needs to cancel an appoint-ment?
- Agreed response time to e-mails and actions that should be taken if either party does not response to e-mails

What mix of media did you use? Which media worked best/least?

Initially it was agreed to engage in weekly telephone contact for the first four weeks and then every two weeks. At the client's request, telephone sessions were supplemented with on-line dialogue to provide an update on actions taken or to challenge whether they had been carried out. As the re-lationship progressed and the client became more self-sufficient, the main contact was periodic e-mail.

E-mails were used by the client to confirm agreed goals, identify issues to be covered in the telephone sessions, and ask questions between telephone sessions. The coach used e-mails to forward diagnostic questionnaires and other material discussed in telephone sessions and also to "nudge" the client if there had been a prolonged period of silence.

A potentially diverse mix of media could have been used through the session: for example, MSN messenger or Skype would have been constructive and enhanced the process.

How did the virtual environment compare with your experience of face-to-face developmental alliances?

The combined use of telephone discussion and e-mails provided a richness that can sometimes be missing from face-to-face discussions. It also provided time for reflection between exchanges as well as allowing for better preparation. Telephone contact supported by e-mail contact built confidence and rapport during the early sessions, and e-mails became the preferred contact method as the client became more self-sufficient.

Both parties involved in this case study were very familiar with communicating via telephone conversations and e-mails. This may not be so with all coaches and clients. An opportunity to meet informally face to face was very useful but not critical in this instance.

Can you give us a flavor of the discussions?

The discussions were focused on the challenges that the client had identified since the last sessions and his progress in working towards his identified goals. Initially the relationship was more that of mentor as the coach had experience in setting up and managing a small business. This enabled pertinent questions to be asked of the client and to make the client aware of potential sources of relevant information, techniques, and sources of support.

What concerns, if any, did you have about the relationship? How were these resolved?

The main concerns experienced by the coach were long periods of silence caused by the client's personal circumstances and work demands. E-mail nudging helped to maintain contact and the relationship.

What were the outcomes? How, if at all, did they differ from the original goals?

At the end of the coaching relationship the client considered that the goal to be challenged on what he was doing had been fully met. Accessing support in channelling his thoughts, setting goals, and taking action to achieve had been mostly met but we considered that further exploration of reasons for not taking action and what to do about it would be worthwhile.

The client also reported that the relationship had given him greater awareness of the experience of being coached, increased knowledge of the contracting side of coaching/mentoring and the importance of asking permission before asking a difficult question or when offering advice or giving feedback.

What lessons did you learn?

Not all contracting issues can or should be dealt with initially as the client may not have sufficient experience of the process to make an appropriate response—for example, asking the client's permission to ask challenging questions but also making it clear that they are not under any obligation to respond. Have a checklist or diagnostic list to ensure that all areas are covered, or, if they are postponed there is a "catch-up list."

E-mails were very effective in following up telephone sessions and exchanging ideas and thoughts.

Some learners do not want to commit to a definite contact arrangement but use the as and when approach. It is therefore important for the coach to be flexible but at the same time ensure good practice is followed.

What would you do differently?

Contracting is critical. It may not be appropriate to cover all issues initially but these do need to be covered during the early sessions. Be aware of what is critical and do not overlook contracting at any stage because of the client's hurry-hurry behavior pattern.

Take time at the beginning of the relationship to clarify client's understanding and expectations of the coaching process. A previous knowledge of coaching can enable the relationship to get off to a positive start. A lack of awareness may signal the need to spend more time on contracting and reviewing at the start of the relationship.

Have a specific contract in place for beginning, reviewing and ending the relationship. Useful questions for review are:

- What has been achieved?
- Is the coaching still adding value?
- Where do we go from here?
- What needs to change?

Unavailability may break the flow. If this should occur either the coach or client should give as much notice as possible. Using e-mail nudges to break silences can be invaluable.

Be prepared to use a range of other media that can be used to improve the effectiveness of an on-line relationship.

Reflection and consolidation of what has been achieved are critical if clients are to have confidence in engaging in further online coaching.

JULIE HAY—SKYPE SUPERVISION

For those unfamiliar with Skype, it is a facility whereby you can hold conversations (as on the telephone) without any cost beyond that of having your normal internet access. You can also add cameras at either end. Conference calls are possible and those without Skype access can dial in on a normal telephone (but have to pay for that call of course).

There are no doubt similar facilities, but Skype is the one I am familiar with—and I liked the alliteration of Skype Supervision. See more details at www.skype.com.

Qualifying as a teaching and supervising transactional analyst specializing in the organizational and educational fields of application has taken me around the world. This started because most of my TA colleagues specialized in psychotherapy, so I was in demand to provide the training and supervision for those who wanted to focus on development rather than cure.

My travels have continued over the years because there are still relatively few organizational TA'ers around the world. Hence, I became the sponsor of students seeking their own TA qualifications in India, Canada, South Africa, New Zealand, Australia, Sweden, the Netherlands, Slovenia and Romania—as well as Devon, Cumbria, and Scotland.

While I've thoroughly enjoyed visiting such a range of countries, it is not realistic to keep visiting each of them every year. In some, we have reached the stage where one of my students has now attained the level where she has taken over my role. In other countries, though, we have not yet reached that stage.

Of course, they get training and supervision from other visiting TA trainers. They also study with TA psychotherapy trainers and we look at how to adjust the therapy approaches accordingly; some of them are able to attend national and international TA conferences, and some travel to (or within) the UK to attend my regular programs.

This still left a gap in the number of supervision hours I could provide. Seeing a range of supervisors is good practice because their approaches and what they pay attention to will vary—but within the TA community it is expected that a set proportion of these are done with one supervisor so that ongoing issues and themes can be picked up.

Taping and Analyzing

Thus, over the years we have developed ways of supervising that rely on technology. Telephone calls were too expensive, and often didn't work in any case in some parts of the world. What we did instead, and still do, is:

1. The supervisee audiotapes (or videotapes) their work—that is, they tape a coaching session or film a training workshop.
2. The supervisee then listens to (or watches) the tape and identifies segments where "something interesting" is happening. Beginning supervisees may pick segments more or less at random, but as they become more experienced, they are able to pick out segments where their interventions, or reactions to clients, are not as potent as they might have been.
3. The supervisee then analyzes the selected segment. This is normal practice for us—we are transactional *analysts,* so we analyze. They take a series of TA constructs in turn and analyze what is happening—for instance, what ego states are involved, what strokes are being exchanged, what windows on the world (life positions) seem to be operating, what discounts (unconscious overlooking) they hear, and so on.
4. The supervisee may also prepare a transcript, especially if they are working in a language other than English. This helps me to follow the dialogue. Note that they don't always do this; providing a transcript of even a few minutes is extremely time-consuming. It does, however, stimulate a lot of skill of analysis and is a requirement for the TA oral exams, so advanced trainees need the practice at analyzing their own work in this way.
5. As the supervisee listens and analyzes, he or she is also thinking about how else he or she might have behaved. This applies whether the intervention made was successful or not; it is all part of creating an increasingly varied array of options to choose from in the moment.
6. The supervisee prepares a covering memo to me that outlines:
   - the overall contract with the client(s)
   - the sessional contract
   - their "diagnosis" of the client(s)
   - what they want me to pay attention to when I listen to the recording
7. I receive tape, memo and possibly transcript. I listen and may respond by sending written feedback or by dictating a recording that they in turn can listen to.
8. We may then exchange further comments if necessary—the aim being always to further develop the professional competence of the supervisee.

The process described above was still a bit remote. We had to rely on occasional meetings to develop our supervisor/supervisee relationship. It also meant that I could not observe their reactions to my feedback, and so could not tailor it to their personalities as much as I would have liked. Fortunately TA emphasizes being in the here-and-now and this helps overcome tendencies for supervisees to adopt Child ego state and perceive the supervisor as Parent.

What Skype Has Added

The emergence of Skype has led to some very useful additional options. We will often still use the same setting-up process, but now I can give the feedback via Skype so we can have a supervisory conversation. I can also talk with several supervisees at the same time, giving them the chance to be in contact with each other even though they may be in different countries. This is particularly beneficial when a supervisee is the only one within a country; it also helps to keep TA cross-cultural.

The other major advance that has come from my using Skype is that I can now conduct live supervision of supervision. For this, the supervisee will arrange to have one of their own trainees with them. They set up audio but not usually camera because it would need high-quality video for me to be able to see both of them, and also because TA analysis can be done very effectively on voice only.

I should perhaps explain that I provide supervision with the emphasis on super-vision—helping the supervisee to become increasingly self-aware and able to step back and create their own meta-perspective of their practice. I do also cover the supportive and normative aspects of supervision as well, of course, but expect to be very much in dialogue with my supervisees as they analyze their own performance.

Once we have the Skype connect and have said hello, clarified what is to follow, and so on, I contract with the supervisee for the supervision I will provide, after which the supervisee contracts with their trainee for the supervision they will provide. TA supervision is customarily run in 20-minute slots (increased to 30 or so if translation is involved). This is long enough for supervisees to gain significant insights, avoids potential self-awareness overload, and ensures those involved don't fall into the trap of filling the available time.

After the 20-minute slot has ended, I then supervise in line with the earlier contract. The trainees will usually stay in the room to hear this, although they know that the procedure is for them to listen but not join in.

There are times when my supervision needs to include a 'message' for the trainee. For example, if the supervisee missed something that could lead to problems for the trainee, I will raise this with the supervisee but will be conscious that this is something that the trainee needs to pick up on. A recent example of this was a trainee who had somehow agreed to work separately with a manager and an employee who worked for that manager but was now being asked by the manager for feedback about said employee—

and the trainee was not yet aware that the manager's apparently reasonable request would lead to a serious breach of confidentiality relating to the subordinate—or a serious argument with the manager about who was paying for the trainee's services.

Because I am not physically there to observe the trainee at times like this, I make explicit at the start of each session that I am relying on:

1. trainees to take responsibility for their own reactions to what gets said
2. trainees to ask for what they need—after my supervision is done—if they have questions or are disturbed in some way
3. the supervisee, or preferably a colleague who is there as an audience or awaiting their turn to be supervised, to observe the trainee's reactions and alert me (again, after the supervision is done) if there appears to be an issue still needing attention.

Benefits of Skype

I apply a TA model from Crossman (1966) to asses whether Skype-based supervision is satisfactory:

- I am able to give permissions to those involved to develop their professional competence.
- I am able to provide protection by raising areas where there is a risk of infringement of professional practice guidelines or ethical codes.
- I am able to be potent enough to provide 'good enough' supervision which models ways of interacting that can appropriately be used also by the supervisee with his or her trainee—and in line with positive parallel process (Hay, 2007), by the trainee with his or her client.

I am also aware that Skype allows me to provide supervision to those in economically disadvantaged areas of the world, and to those who are in the forefront of this sort of developmental activity in their country and might otherwise lack professional contact with their peers around the world.

ANJI MARYCHURCH—COACH VIEW

How and why did this relationship come about?
N contacted me having spoken to another coach who referred him, as he recognized that the client needed some career-specific help. We connected well on our first conversation and agreed to a no-obligation discussion before Christmas to discuss his career concerns and establish whether I could help him by "mapping out" what surfaced.

At this time we both had an opportunity to assess the quality of our rapport and decide whether to proceed. N wanted help to stop "spinning the wheels" and find some future career satisfaction and direction. He also recognized that he needed a plan of action to help him make some changes.

What was the initial contract? Did it change over time? If so, how and why?

Our initial contract was for three months, which became spread over four due to some existing commitments the client had already. We are now in touch once a month.

He completed all initial paperwork very quickly and began a dialogue with me by email prior to our planned telecoaching sessions in January. He also sent me a prodigious amount of information about himself and his work projects over the years and took it upon himself to look into several sources of reading plus an online resource I had recommended, so he was keen to get started. I did not refer to this information prior to commencing our relationship as I prefer to form my own understanding of the client and his desires rather than having any preconceptions.

Our relationship deepened over time as I got to know my client and understand the complexity of the issues he brought and their impact on his career and life.

What mix of media did you use?

We used predominantly telecoaching with email support and one face-to-face meeting for some skills analysis. There was some reliance on the internet for books and on-line resources as appropriate to support our work.

Which media worked best/least?

I think the telecoaching sessions plus regular email support really encouraged my client very well. The face-to-face meeting was comprehensive, being twice as long; however, it was not as time- and cost-effective.

How did the virtual environment compare with your experience of face-to-face developmental alliances?

My experience of working virtually with this client was that the sessions were far more intimate and the level of disclosure higher. The potential for distractions such as refreshments and social small talk was greater working face to face and could have impacted negatively on the session outcome without vigilance and focus by the coach.

Can you give us a flavor of the discussions?

Some sessions were very practical and task-orientated in contrast to some discussion threads that evolved into deeper, spiritually based conversations. All produced a variety of actions the client committed to take to increase his self-knowledge and awareness, thereby deepening his learning to facilitate change and growth. I found my client to be an intelligent, courageous person who was willing to do the internal work required to move forward and keep going.

What concerns, if any, did you have about the relationship?
How were these resolved?

My initial concern was that the client was expending a lot of energy on non- productive repetition of his story. I conveyed the strength of the "hold" his story had on him and suggested different ways he could look at himself and his past in order to create new choices in the present. Clues to the future can come from the past, but only changes in the present will determine future successes. My other concern was to encourage him to try different ways of managing his emotions in his personal relationships at work and at play. This encompassed new ways of behaving both one on one and in groups and to practice a range of new activities/habits that would support him going forward. This necessary work on personal issues became the key area of focus in our early sessions to help the client to stop "spinning the wheels" and move on.

What were the outcomes? What lessons did you learn?

My client really grasped the relevance of working on himself first to declutter and refocus his energy before tackling longer-term career changes. He greatly appreciated being listened to deeply without judgement. The power of this enabled him to tackle longstanding personal issues and make forward progress. The value of good quality direct communication was validated frequently and I really enjoyed the opportunity of working with a challenging professional that this client provided. My client also reported increased self awareness and personal confidence with improved clarity about the kind of activities he was passionate about doing. We identified several research tracks for independent exploration.

How, if at all, did they differ from the original goals?

I am delighted that my client embraced the depth offered and responded to the flexibility of approaches to thorny issues! I hope that this will now sustain him as he begins to work from a place of solid foundation. I am looking forward to learning about the challenges coming up for him and hope to continue to be able to guide him wisely on the choices which will emerge as he progresses in his careere/vocational explorations.

What would you do differently?

Nothing!

ANJI MARYCHURCH—CLIENT VIEW

How and why did this relationship come about?

I was chronically dissatisfied with my career/vocation and had tried all the approaches that I could muster. A colleague suggested to me that I should seek out an expert to help me. I called a local coach from the yellow pages who listened to the story and recommended two numbers. I called both and went with A because she connected with me, understood and exuded experience, and had a manner that made me trust her.

What was the initial contract? Did it change over time?
If so, how and why?

We agreed to four hours per month for three months as an initial position. This involved quite intense work, which seemed to come to a natural end and left me in a new phase requiring a more external and exploratory orientation in a much more undirected, emergent way. It made sense to reduce the sessions to one per month to keep contact, encouragement and to "hold me to the task."

What mix of media did you use?

Predominantly we have used telephone calls. We had one face-to-face meeting and have had many emails—some brief, some long.

Which media worked best/least?

I think there is a synergy in having a mix of media. It is not necessary or practical to meet physically every time, although it is helpful to do so to establish a deeper rapport and relationship. Email has been very helpful in providing an efficient way to communicate, receive suggestions, and provide feedback.

How did the virtual environment compare with your experience
of face-to-face developmental alliances?

We could have done it all virtually, but I think there is still a value in meeting face to face to establish rapport and chemistry. There is no getting away from the fact that much of our communication is physical and that meeting face to face is a deeper experience. Having said that, keeping communication at a distance can allow for a deeper level of self-disclosure and intimacy. It's easier to say intimate things over the phone or by email than in person.

Can you give us a flavor of the discussions?

Our discussions have varied enormously. Sometimes they have been "history taking" and "diagnosis," and sometimes very directive—with specific tasks to do. Sometimes nurturing, reframing, encouraging, clarifying, testing; sometimes indulgent, sometimes challenging. The tone has always been non-judgmental and nurturing but honest and direct.

What concerns, if any, did you have about the relationship?
How were these resolved?
   None.

What were the outcomes? What lessons did you learn?
   After three months, I had a clearer idea of my own inner compass—my values and instincts—and was more connected with my heart/gut/body. I was more aware of and more able to live in the present, now, mindfully. I had put to bed some of my hang-ups and past psychological baggage. I am better able to appreciate my successes in the past and to recognize the causes of my failures. I feel ready to break with the past, to redefine myself in all sorts of ways, and to be a new person. I feel I have broken out of some stuck patterns. I am in a fitter and more adaptive position to "get back in the saddle" and have another go.

How, if at all, did they differ from the original goals?
   I expected to end up with a clear strategy and a plan. That would be nice but is unrealistic at this stage. I did expect A to quickly get into strategy and detail without understanding the deep personal stuff, so it was a really pleasant surprise that not only did she get in at that level but did so in a much deeper and more original way than I had been able to do so alone. It means that whatever follows is properly grounded. This hopefully will reduce the risk of being derailed.

What would you do differently?
   Nothing.

REFERENCES

Crossman, P. (1966). Permission and Protection. *Transactional Analysis Bulletin 5(19)*, 152–154.

Hay, J. (2007). *Reflective practice and supervision for coaches*. Berkshire, UK: Open University Press.

CONTRIBUTORS

Lisa A. Boyce is the director of Behavioral Science Information Technology Applications Research at the United States Air Force Academy. Lisa's research focuses on leadership training and development, including leadership coaching, leader self-development, and military leadership with a particular interest towards the application of technology. She has over 50 research publications and presentations on related topics. Lisa is a member of Academy of Management, the American Psychological Association, Society of Industrial and Organizational Psychology, and Military Psychology.

David Clutterbuck (Ed.) is visiting professor of coaching and mentoring at both Sheffield Hallam and Oxford Brookes Universities. He is the author, co-author, or co-editor of 14 other books on coaching and mentoring, amongst a total of nearly 50 books all told. David founded Clutterbuck Associates, an international provider of consultancy and support for mentoring programs. He led the development of the International Standards for Mentoring Programmes in Employment and co-founded the European Mentoring Centre, which evolved into the European Mentoring and Coaching Council. He has assisted hundreds of organizations around the world in designing and sustaining programs of coaching and mentoring.

Randy Emelo is President and CEO of Triple Creek Associates and has over 20 years of experience in management, training, and leadership development. He has worked with military, for-profit, and nonprofit organizations, often focusing on leadership development and learning initiatives. A prolific author, Randy writes Triple Creek's monthly newsletter on mentoring that reaches 21,000 subscribers and has authored numerous articles and research papers on topics related to mentoring, collaboration, innovation and creativity, management practices, and leadership development for Tri-

Virtual Coach, Virtual Mentor, pages 241–244
Copyright © 2010 by Information Age Publishing
All rights of reproduction in any form reserved.

ple Creek and such publications as *T+D*, *Chief Learning Officer*, and *Learning Circuits*. He has also given numerous talks and presentations on such topics to varied audiences. Randy holds a master's degree in organizational design and effectiveness from the Fielding Institute in Santa Barbara, CA.

Kathy Froud is a qualified coach-mentor, a Chartered Fellow of the CIPD, and a member of the Chartered Institute of Management and the Institute for Learning, supporting and enabling the learning and development of professionals has been the focus of her career. She currently runs her own learning and development consultancy, cpdservices@yahoo.co.uk.

Julie Hay is internationally accredited as a TA supervisor and trainer and internationally licensed as an NLP trainer. She was a founding member and President 2006–2008 of the European Mentoring & Coaching Council. She has many years of managerial and consultancy experience in large organizations; is the author of several books, audiotapes, and packs (TA, NLP, coach/mentoring, assessment & development centers); and has recently set up the International Centre for Developmental Super-Vision, www.icdsv.net.

Gina Hernez-Broome is a Senior Faculty member at the Center for Creative Leadership's Colorado Springs campus. She is lead faculty and program manager for both the Leadership Development for Human Resource Professionals (LDHRP) and the Coaching for Human Resource Professionals (CHRP) programs. Additionally, Gina is project manager and lead researcher for the Center's coaching research and evaluation efforts. She has presented her work at a variety of conference, including those sponsored by the Society for Industrial and Organizational Psychology, Academy of Management, and the International Coaching Federation.

Zulfi Hussain MBE is former chair of the UK branch of the European Mentoring and Coaching Council. Originally a specialist in computer security, he initiated the e-mentoring programs within British Telecom, which was one of the first organizations to approach mentoring in this way. He is one of Yorkshire & Humber's best-known and most respected entrepreneurs and has been recognized as one of the top 100 most influential people in the region.

Irja Leppisaari is principal lecturer of online pedagogy at AVERKO (the Open Online University of Applied Sciences) administered by the Central Ostrobothnia University of Applied Sciences. Her post-doctoral research focuses especially on e-mentoring, and the professional development of both educational staff and workplace personnel. During the last several years, she has presented research papers at numerous international conferences. Dr Leppisaari is actively involved in online pedagogical research and development at the Finnish Online University of Applied Sciences. She is a sought-after expert in various online teaching and learning and e-mentoring development and evaluation ventures.

Rusty Livstock is Deputy Chief Executive at "LTL Connect the online coaching company." Rusty has spent the past 20 years working on the human dimensions of workplace profitability. She spent 12 years leading resilience services in a large, manufacturing organization which took her across much of the globe, including periods employed in Germany and China South and destinations as varied as South Africa, Kuwait and Isarel. As well as an MBA she is coaching and counselling qualified.

Tom McGee is V.P. of Special Projects for Triple Creek Associates, where he oversees research on web-based mentoring and leads writing projects for Triple Creek's many white papers and booklets on mentoring that reach 19,000 subscribers. He also directs the research, expansion, and delivery of mentoring training programs. With over 20 years of leadership and consulting experience, Tom continues to be a lifelong learner and studies extensively in the areas of organizational development theory and leadership theory and practice. He holds degrees from Texas A&M University and Dallas Theological Seminary, and he has completed doctoral research in the area of mentoring.

Dr. Angus Macleod is author of *Performance Coaching* and *Me, Myself, My Team* (both by Crown House) as well as *Self-Coaching Leadership* (John Wiley) and a multitude of papers on coaching in the international press. He is a regular contributor to conferences and facilitates master-classes in coaching around the world as well as coaching 1–2–1 internationally. He designed two phenomenally popular Diploma courses for Newcastle Colleg, providing many thousands with coach training since 2004. His organization, Angus McLeod Associates, trains coaches and managers as well as supplying coaches to organizations. www.angusmcleod.com

Anji MaryChurch is a learning and development practitioner with 30 years experience bridging both business and education, of which the last thirteen years have been in coaching and mentoring, with career development being a central theme. She has been coaching corporate individuals and groups on career transition issues, including outplacement since 1995, and has a thriving career coaching practice. Anji is one of a small number of people in the UK who hold the credential of Professional Certified Coach (PCC) awarded by the International Coaching Federation (ICF).

Edna Murdoch is a Director of the Coaching Supervision Academy and is an Executive Coach and a Coach Supervisor. She has supervised Executive Coaches, Senior Managers, and Psychotherapists since 1991. She has trained in both Clinical and Transpersonal Supervision, and currently trains coaches in Coaching Supervision through CSA. She runs live and telephone coach supervision groups and supervises executive and life coaches. www.coachingsupervisionacademy.com

Kim Rickard is an e-mentoring researcher and practitioner. She has delivered an annual program for professionals operating as independent contractors and consultants since 2002, and at time of writing was completing a PhD that considers the effectiveness of e-mentoring in the small business context.

Jean Simpson is the Sciences Student Guidance Advisor at Staffordshire University.

Paul Stokes is the Deputy Director of the Coaching & Mentoring Research Unit (C&MRU) at Sheffield Hallam University and is Course Director for the MSc in Coaching & Mentoring courses, both in the UK and in Switzerland. He is an active member of the European Mentoring & Coaching Council, serving on the European Conference Committees and the UK Conference and Research Committees. As well as being an independent coach and consultant in his own right, Paul is an active researcher and academic, having co-written two books and several book chapters and academic journal articles on mentoring and coaching. His current research interests include "the skilled coachee" and the role of supervision within coaching and mentoring.

Leena Vainio is the head of HAMK eLearning Centre in HAMK University of Applied Sciences. Her research focuses especially on organizational change management and leadership when e-learning methods are used in teaching, learning and working. She has more than 25 years of experience in the learning design and development fields. She has presented at many conferences, and has written many articles, and book chapters concerning e-learning, e-mentoring, e-working and leadership. Vainio has been eight years the chair of the Finnish eLearning Association.

Janet Wright is currently a principal lecturer and the subject leader for Geography at Staffordshire University. She is the project leader for the Faculty of Sciences' e-mentoring scheme, which has been running at Staffordshire University since September 2006.

The Brightside Trust would like to acknowledge the following contributors: Ian Akers, Development Manager, Big Deal Blogs; Tracy Sacks, Project Manager, Bright Journals & Big Deal Blogs; Suzanne Maskrey, Senior Project Manager, Live Journals; and Viki Nicholson, Senior Project Manager. info@thebrightsidetrust.org

LaVergne, TN USA
25 January 2011
213871LV00002B/46/P